Planning for the Future

Felicitas Selter

Planning for the Future

Do Animals Have a Time-Relative Interest in Continuing to Live?

mentis

Gedruckt mit Unterstützung des Förderungsfonds Wissenschaft der VG WORT

Umschlagabbildung:
Foto eines Florida-Buschhähers

Bibliografische Information der Deutschen Nationalbibliothek

Die Deutsche Nationalbibliothek verzeichnet diese Publikation in der Deutschen Nationalbibliografie; detaillierte bibliografische Daten sind im Internet über http://dnb.d-nb.de abrufbar.

Zugl. Dissertation an der Graduate School of Systemic Neurosciences der Ludwig-Maximilians-Universität München 2018

(Koninklijke Brill NV, Leiden, Niederlande; Brill USA Inc., Boston MA, USA; Brill Asia Pte Ltd, Singapore; Brill Deutschland GmbH, Paderborn, Deutschland)

www.mentis.de

Einbandgestaltung: Anna Braungart, Tübingen
Herstellung: Brill Deutschland GmbH, Paderborn

ISBN 978-3-95743-177-6 (paperback)
ISBN 978-3-95743-736-5 (e-book)

Acknowledgments

There are many people who have contributed to this thesis. First of all, I want to thank my supervisor Stephan Sellmaier for his continuous encouragement, enthusiasm and guidance. He put me back on track whenever I went astray in my own thoughts.

I would also like to thank my two other advisors, Markus Wild and Benedikt Grothe, whose helpful comments on earlier manuscripts made important contributions to this dissertation. I also want to thank external reviewer Christoph Hoerl for his time and effort.

Current and past members of the *Research Center for Neurophilosophy* have shared their thoughts, insights and ideas with me in our colloquia and provided helpful feedback. I want to particularly thank Steffen Steinert, Joachim Lipski, Janett Triskiel, David Kaufmann, Lara Pourabdolrahim, Michael von Grundherr, Mario Günther, Cameron Beebe, Ali Yousefi Heris, Benedetta Romano, Gasper Stukelj, Sachith Cheruvatur and Urim Retkoceri.

Moreover, I would like to thank all the speakers and participants in the workshop *The Self in Time*, whose contributions made those three days such a great experience. My deep gratitude also goes to Jeff McMahan, who took the time to personally discuss and clarify some of his ideas and arguments and encouraged me to pursue my argument.

I want to thank the staff of the *Graduate School of Systemic Neurosciences* for providing an academically and personally stimulating atmosphere and for financially supporting my PhD project with a generous stipend. I also would like to thank the *mentis Verlag* for publishing my thesis and Michael Kienecker for his helpful advice on several occasions.

In a very real sense, this thesis would not have been possible without my family and friends. A special thanks goes to my parents for their continuous support and love. John took the time to proofread my manuscript and did a great job. I also want to particularly thank Barbara, who whole-heartedly encouraged me and once in a while contributed the helpful reminder that there is more in life than a PhD.

Finally, I want to thank Patrick. To say that the last couple of years have been challenging to both of us – with at least two black dogs having come into our lives –, would be an understatement. I am deeply grateful for his strength, patience and love.

Table of Contents

CHAPTER 1

Introduction

Discussions about the moral permissibility of killing animals commonly focus on the pain and suffering accompanying the killing. Many people are opposed to the practices of animal agriculture, for instance, because of the distressful conditions under which the animals are raised, kept and finally slaughtered. Fortunately, the times that the reality of animal suffering were challenged, are long-gone. There is a general consensus nowadays that animals[1] can be put in a state of pain, distress or boredom and have an interest in not being made to suffer, therefore. But assume that it would be possible to raise and kill animals for food and experiments in ways that caused them absolutely no pain or suffering. Would it still be wrong to end their lives?

Take the example of the young pig Edward, who happily lives on a small family farm. Is the farmer morally wrong in painlessly killing Edward and eating his meat? Assuming that Edward could live for at least another five, overall pleasant years. Even if it is taken as a given that pigs do not have a concept of life and death, these five years surely have some weight. But for whom? Who cares about these future goods the animal will be deprived of?

Right away, this thought experiment brings us into the research topic that this thesis deals with, that is, whether and to what extent animals are harmed by their untimely deaths and, thus, whether they have an interest in their continued existence. As I understand the term, a *premature* or *untimely death* indicates that life has not yet come to completion or full development, because, e.g., the deceased was young or still had plans for the future. In case of animals, it might refer to missed chances for pleasure-maximization, such as the fact that we pity a dolphin, who has never swum in the ocean, or a pig, a highly social animal, who is confined to a solitary life in a narrow cage. Even though not everyone would probably agree that an animal life can be incomplete in this sense, it should be uncontroversial that animals can die untimely deaths, if untimely is defined in terms of life span. I will not discuss the exact point in time that defines an untimely death, but assume that it lies significantly above the average life expectancy, but not over what is biologically possible for representatives of the said species. Thus understood, a death will be premature

1 For reasons of readability, I will mainly focus on animals with intermediate cognitive capabilities, e.g., dogs, pigs and fish, but not insects. Whenever the term *animal* will also be used to include those animals with higher cognitive capabilities, it will be evident from the context.

 | DOI:10.30965/9783957437365_002

or untimely in most cases. It should be noted that this does not yet make any assumptions about the harmfulness of a particular death. In some occasions, it might be in the interest of a subject to die rather than to continue living, either because personal completion has been achieved, which would make death 'mature', or because future prospects are very poor. A young individual can thus have an interest in dying and still die an untimely death.

To many readers, the question whether and to what extent animals have an interest in their continued existence will undoubtedly sound rather absurd at first glance, but not necessarily for the same reasons. Whereas some might take it as rather self-evident that every sentient animal loses something from dying, others might disagree in pointing out that animals do not even know that they will continue to exist tomorrow, so it would be absurd to ascribe them with an interest in continuing to live. Animals are frequently said to be much more anchored in the present, living their mental lives in the here and now. Human persons[2], in contrast, are typically extremely future-oriented. Because animals are stuck in the present, it could be argued, they are not aware of having a future and cannot be deprived of something they do not even know they have. Consequently, animals cannot be ascribed with an interest in continuing to live. And yet, death would deprive the pig Edward of roughly five overall happy years. Should that fact alone not be enough to say that he has an interest in continuing to live?

Philosophy has traditionally moved between these two extremes. Albert Schweitzer, on the one hand, was convinced that all living things are subdued under an equally strong desire to live (Schweitzer 2008, 2011). Jeremy Bentham, on the other hand, whose famous footnote in the *Introduction to the Principles of Morals and Legislation* is still frequently cited in the animal rights movement,

2 Which kind of beings qualify as persons has been vigorously discussed not only among animal ethicists for a long time. Here and hereafter the term *person* will refer to those living beings fulfilling certain criteria that are commonly taken to be relevant, such as self-consciousness, rationality and autonomy, among others. That these concepts, all of which are matters of degree, are defined rather vague is intentional in order to account for slightly different understandings of the term. Peter Singer's definition, e.g., is not congruent with Jeff McMahan's concept of a person, as will become evident. To a certain extent, the concept of a person will be used dynamically and flexibly in this thesis, consequently. Still, the generally accepted case of a typical person across most moral theories remains to be a paradigmatic human adult. On the same time, the concept of a person is neither exclusively limited to members of the species *homo sapiens*, nor are all human beings persons. An exact definition is not of vital importance, for the time being, but will become significant in chapter 6. Until then, the term will be used broadly, including most human adults and older children as well as some animals. When specifically referring to human persons, these will also be addressed as paradigmatic human adults to avoid such unfortunate vocabulary as 'normal' or 'healthy'.

did not believe that the killing of animals – always under the premise that death occurs absolutely painlessly – poses a moral problem whatsoever, for animals "are never the worse for being dead" (Bentham 1789/1970, 282f fn b). Jeff McMahan's position can be seen as an attempt to find a middle ground between these two extreme positions and to reconcile some of the opposing intuitions concerning animal death that many of us share. Before coming back to these issues in the first part, a few words will be said about some of the main concepts of this thesis. For convenience, these concepts will be explained by reference to persons, not animals.

Most agree that killing a human person is *ceteris paribus* one of the most fundamental moral wrongs. If the focus is not on the murderer and the bad consequences this act might have on her character or well-being, but on the victim herself, most of us also agree that killing is, above all, a moral wrong, because the victim is *harmed* by her untimely death. However, we also tend to evaluate a certain death's harmfulness according to external factors, such as third party interests or thoughts concerning social utility. The fact that the deceased leaves three small children behind or that she alone could execute a highly complex neurosurgical technique that would have saved hundreds of lives, might contribute to our evaluation of her death. However, in my definition, they leave questions of *harm* done to the dying untouched.[3] One could say that one particular death becomes more *tragic*, the more it is mourned or the more third-party interests are violated by it, but it does not automatically contribute to its harmfulness. What I am interested in is the harm that death constitutes for the subject herself.

An event is commonly considered to be potentially harmful in two different ways. Either, an event is intrinsically bad or leads to something that is bad for the individual itself, or an event can deprive the being of something that would have been good for her. As death is not intrinsically bad, it can only be bad in the second way.[4]

---

3 It may well be that her children's well-being is the mother's greatest wish in life, or that the neurosurgeon has worked her whole life towards the application of her life-saving technique. In these examples, third-party interests might fall together with very strong desires the deceased themselves held. Whether the thwarting of categorical desires contributes to or even constitutes the harm done by death will be discussed in chapter 2.

4 *Bad* and *harm* will be used interchangeably, first and foremost, because Jeff McMahan, whose account represents the moral fundament of this thesis, does not differentiate between the two as well. Harm will also be restricted to living beings with a welfare, i.e., the term will not be similar to other, for instance, jurisdictional, usages of the terms, in which harms can be done to cars, reputations or the environment (for a contrasting view to McMahan's definitions of bad and harm, see Harman 2011).

Whether a harm also necessitates a consciously experienced negative mental state is a much disputed issue (for an extensive overview, see Bradley et al. 2013). David Suits argues in a variety of essays that because death does not negatively influence, but simply *eliminates* a being's welfare, it is logically impossible to define it as a harm. It is however equally possible to understand the irreversible ending of consciousness in itself as negatively affecting the welfare of a being, precisely because it deprives her of the opportunity to ever experience anything again.[5] Death, on this view, is not to be equated with a vanishing of well-being (similar to a desire that just vanishes one day, unnoticed by its holder) but as an elimination of the opportunity to have well-being at all. Death, it will be assumed, does in fact do a harm, due to its privative nature.

Another presupposition that will be made, but not defended, is that death's harm comes in degrees. This means that some deaths can be worse than others, both in the life of one and the same person at two different points in time t_1 and t_2 (*intra-subjective relation*), as well as the death of subject A compared to the death of subject B (*inter-subjective relation*). As the first claim is not very controversial – in most cases, death is deemed to be worse the earlier in life it occurs, for example – I will take the first assumption for granted without arguing for it. Yet, if we do accept that harm might differ intra-subjectively, we should also accept that it can differ inter-subjectively. It quickly becomes apparent, however, that this latter claim is a bit more complicated, because there seem to be contrasting intuitions[6] around, which can be illustrated nicely by the help of Tom Regan's notorious lifeboat case (Regan 2004, 285).

In this thought experiment, five survivors are on a lifeboat, four human persons and one dog. The boat can only support four passengers, so one individual has to be thrown over board. Death will in all cases occur instantly and painlessly. If the survivors refuse to make a decision, all of them will die. Who should be sacrificed? I stipulate – just like Regan does – that everyone would agree that the dog has to go. But while some people, including me, will reason that death does less harm to the dog than it does to any of the four persons,

---

5 This argument has been put forward in McMahan (1988), even though McMahan himself argues against it on the grounds that experience is not required for something to be harmful.

6 We all have certain intuitions about the comparative badness of death, some of which are strongly held (e.g., that it is generally worse to die young). A plausible theory of death's harm should take these intuitions seriously and not abandon them lightly. With respect to such a complex issue like the morality of killing we should be well advised to rely on our intuitions – but only to the extent that they do not clash with other deep rooted intuitions, they do not lead to implausible implications and, importantly, they can be built on stable argumentative ground. Intuitions will serve as a touchstone, so to speak, but they are not sacrosanct (see also chapter 6).

some might be inclined to say that actually there is no moral reason whatsoever to sacrifice the dog. Every living thing is completely subordinated under its biological will to survive. Death represents the *greatest possible loss* any living being can suffer. For each one of the five individuals life is all it has and is therefore equally valuable. Differences in complexity, intelligence, and, above all, consciousness are insignificant. This is the position Albert Schweitzer famously held.[7]

Yet, why should the biological will to survive impose any ethical obligations on us? As Stefan Eck points out, such a position is in grave danger of making a naturalistic fallacy. Schweitzer just assumes, without arguing for it, that the biological will to survive in any living organism is analogous to a human person's will (or desire) to live and therefore is worthy of the same respect (Eck 2002, 51f). This can, to say the least, be questioned. To avoid this criticism, one would have to find a better argument why death harms all living beings equally.

One possibility could be to argue that we will never be able to know what death means to, say, a frog, so we simply are not in the position to make any cross-species comparisons and should better err on the side of caution in pretending *as if* death were equally harmful to all living beings. I agree insofar that cross-species comparisons are challenging, but disagree in that they are unsolvable without inevitably falling into anthropocentrism. Anthropocentrism in this context refers to the view that death can only be harmful as long as it is being suffered by a human being or that the more alike animal and human are, the more harm death does to the animal. Most people strongly believe that a frog is not equally harmed by his death as a cow is, anyway. However, someone

7 Alternatively, one could argue that there probably is no objective reason for sacrificing the dog, i.e. acknowledging that we are indeed biased in choosing the dog, but simultaneously argue that we have a *moral duty* to do so, anyway. This does not need to be a contradiction if moral reasons are allowed to be derived from two different sources. Most of the time we refer to intrinsic properties of the being that will be affected by our action, but sometimes relational facts are taken into account as well. In the lifeboat case, the human passengers would thus refuse to throw one of them overboard for reasons of "justified partiality" (DeGrazia 2007; McMahan 2002, 218). Yet, many authors have extensively, and, in my opinion, rightly, argued against the moral relevance of biological species alone, so these arguments will not be ruminated here. Alternatively, it has been argued that not the biological features matter but the special relation that connects one human individual with all other human beings (e.g., Nozick 1983, cited in Cavalieri 2002, 95). Just like a father who is justified (or even morally obligated) to save his child rather than a stranger's child from a burning house, so is the human passenger obligated to sacrifice the animal and not one of his kind. Whether co-membership in a biological species is such a special relation and might be (partly) morally relevant, is a tough question. It should be emphasized that to deduce moral claims from biological features is highly problematic in all cases, because it invites analogies to racism or sexism (for an extensive discussion of speciesism, see, e.g., Cavalieri 2002 and Pluhar 1995).

devoutly believing that death is equally harmful to animals, humans and other living beings alike will not be persuaded from the contrary by the considerations put forward in this thesis.

Put together, two presuppositions have been established. It will be assumed that death *ceteris paribus* does a harm to the one who dies and this harm will be regarded to be a matter of degree. Just as one death at a particular point in time in the life of one subject can be worse than at another point in time, one death of one individual can be considered to be worse than another individual's death.

What is it exactly that death takes from us? And how do we make sense of the intuitional claim that some deaths can be worse than others? In the second chapter, I will suggest that Jeff McMahan's *time-relative interest account* (TRIA) offers itself as an especially promising candidate to give an answer to these questions. In doing so, I will point towards certain major flaws the two most common alternative positions face, and argue that the TRIA in fact combines the strengths of both approaches without including their deficiencies.

The third chapter will provide an analysis of the TRIA and argue that all sentient animals have an interest in continuing to live and are therefore being harmed through their untimely deaths. However, animals can only form a strong time-relative interest in continuing to live, if they are capable of forming a particular kind of direct psychological connections into their futures. Planning lends itself to empirically investigate these internal reference-typed connections in animals.

In chapter four, I will argue that planning and other related terms have loomed large in the debate around future thinking capabilities of animals, but suffer from a lack of conceptual clarity. I will propose to separate anticipatory planning from strategic planning. Drawing on behavioral evidence in the following chapter, the case will be made that while there is evidence for some animal species (most notably great apes and corvids) indeed possessing anticipatory planning abilities, the evidence for strategic planning thus far is scarce. Strategic planning is relevant for McMahan's account insofar that it binds an individual stronger to her future than anticipatory planning does.

Finally, in the sixth chapter, I will come back to the question, how much moral relevance the animal's interest in continuing to live should be granted. I will argue that some major animal ethical theories do not or only deficiently accommodate the fact that future-oriented cognition is complex and multifaceted. I will also argue that the argument from marginal cases is only of limited use in this context. Neither are animals stuck in time, nor are they full-fletched mental time travelers.

In the final conclusion, I will consequently propose to classify animals into (at least) four groups with regard to their future thinking capabilities. According to the TRIA, the members of these groups have increasing moral relevance regarding their lives. Based on the hypothesis that a time-relative interest in continuing to live of a certain strength can justify the ascription of a strong right not to be killed, which can only be violated in very exceptional cases, it will be proposed that animals with planning abilities should be ascribed with such a right. In my concluding remarks, I will sketch some practical implications that would follow from an acknowledgment of the above considerations with regard to our exploitive relationship with animals.

CHAPTER 2

The Privative Badness of Death

2.1 The Thwarting of Desires

Death, even if it is not accompanied by any negative feelings like pain, fear or regret but occurs instantly and painlessly, often thwarts the individual's desire[8] to continue living. Most of us do not *want* to die. Death violates this desire, therefore. Even though this points towards a very important aspect, the badness of death cannot be explained solely via the desire to continue living.

For one thing, one does not always have to explicitly express a desire for x for x to be in one's interest and, accordingly, to be harmed by x's non-occurrence. In everyday language, interest has two different meanings; one can *be interested in something* and something can be *in one's interest*. In accordance with Regan, the first group of interests will be called *preference interests* and the latter *welfare interests* (Regan 2004, 87f). To have a welfare interest in something (and therefore to be harmed by its non-occurrence) does not require having a desire for it. Worse, even in rational beings, the existence of a desire does not mean that its satisfaction is always in the welfare interest of the person. One can have bad desires, as in the case of drug addictions, or uninformed desires, like Romeo had, who surely would not have longed for his death, had he known that Julia was still alive (DeGrazia 1996, 215).

One could, however, defend the claim that it is not the specific desire to stay alive but the capability to want something in the first place that gives rise to my being harmed by death. By saying 'I want to stay alive', I am not so much referring to the explicit desire to continue living, but rather to the many *categorical desires*[9] that will be frustrated. This way, Bernard Williams famously argued, "wanting something itself gives one a reason for avoiding death" (Williams

8 Throughout this thesis, it will be assumed that animals are, in principle, capable of having beliefs, desires, needs and interests. Particularly the claim that animals can form beliefs and desires has been frequently disputed on the grounds that having beliefs and desires requires the possession of language (see, e.g., Davidson 1982). A response to these skeptical claims lies well beyond the scope of this work, however. Having desires will be rather minimally understood as "having dispositions to act" (Schroeder 2009) or as being in a motivational state, i.e., a state with the function to bring about changes in the environment (Wild 2010, 123ff).

9 William's conceptual discrimination between conditional and categorical desires is widely used, albeit in very different interpretations, due to Williams' own rather vague definition. It has been subjected to a number of criticisms, consequently (Bradley and McDaniel 2013; Bradley 2015).

 | DOI:10.30965/9783957437365_003

1973, 85). Categorical desires are desires that are not conditional on being alive, which is generally understood to mean that the desire would not be cancelled, if the desirer were to die (Bradley 2015, 54). Long-term plans, goals and projects to be achieved in life are common examples for these categorical desires. Whether my *conditional desire* – conditional on being alive – to scratch an itch will be satisfied or simply vanishes, in contrast, does not make a difference to me.

Someone could now defend the view that death does a harm to the one who dies, if it frustrates someone's categorical desires, because categorical desires are desires "for which continuing to live is necessary and which provide reasons for [one's] living on" (Belshaw 2015, 37). Categorical desires, on this reading, both motivate and justify one's continuing to live (Bradley 2015, 54). Christopher Belshaw (as well as Jeff McMahan[10]) now claims that fetuses, infants and most animals are not capable to form categorical desires. While Belshaw draws the conclusion that they cannot be harmed by their deaths in a way that morally matters, McMahan puts much less emphasis on these individuals' assumed lack of categorical desires, as will be seen below.

There are several problems with desire-based accounts like Belshaw's, attempting to entirely capture death's badness by reference to the desires death thwarts, that suffice to reject them. Most importantly, they cannot account for the vast future goods being undesired at the moment death occurs, be it because the person is indifferent or even hostile to those future losses[11] or because the subject does not (yet) have the cognitive abilities to form desires at all. Although it may well be plausible to assume that the harm done is *greater*, if the future good was desired by the individual, and also that the harm is

10 McMahan argues that a categorical desire provides a reason for continuing to live to ensure its satisfaction, "and thus in a self-conscious being [i.e., a person] naturally gives rise to a second-order desire for the continuation and satisfaction of the first-order desire" (McMahan 1992, 3f). On the first glance, it is not clear if McMahan proposes that the possession of second-order desires is merely a side-effect, so that non-persons could, in principle, have categorical desires as well, or if he takes the capability to form second-order desires to be constitutive for categorical desires. I suggest the latter, however, because he excludes non-persons from being capable of forming categorical desires.

11 In order to avoid the unwelcome implication that an anesthetized or severely depressed individual without any categorical desires, currently robbed of her future-directed plans and projects as well of her desire to continue living, would not be harmed by her death, Belshaw claims that death is not only bad for those, who presently have categorical desires, but also for those, whose categorical desires "are temporarily in abeyance" (Belshaw 2015, 38). Belshaw seemingly wants to argue that the fact that the individual will have categorical desires in the future is sufficient to state that she would be harmed even now by dying.

greater the *stronger* the individual desired it beforehand, the fact that the subject would have considered this future good as valuable when it would have occurred should be enough to consider it as a loss the subject suffered.

To meet this objection, the desire-based account could be modified in two ways. Either, one could include not only the prevention of the satisfaction of desires but also of the formation of desires (McMahan 1992, 12f). Death, then, would be bad, because it deprives a subject from the satisfaction of her desires and because it deprives a subject from the formation of all of her future desires. This, however, is very similar to a position stating that it is the loss of future good (be it currently desired or not) that makes death bad, so I will not go into detail of this view here, but discuss it below. Note, however, that a thus modified desire-based account would still exclude small infants, many animals and severely mentally disabled humans from death's harmfulness, at least if it entirely focuses on categorical desires in the way defined above, that is, as long-term plans, life goals and so on.

Alternatively, one could revise the desire-based account to the effect that it is not relevant if the subject *actually* had the desire to stay alive, but if she *would have* that desire under ideal conditions. Depending on how broad these ideal conditions are to be understood, one could even grant that infants, animals and cognitively disabled humans are harmed by their deaths and thus avoid the above raised objection against desire-based accounts.

As Thomas Scanlon argues, however, such a modified account should no longer be regarded as a desire-based theory, because it dissolves into hedonism. Why is it good at all that a desire is being satisfied? Either, because the resulting good is considered to be a substantive good (as objectively good), or because it contributes to the individual's level of happiness and well-being (as subjectively good) (Scanlon 1993, 186–192). To determine if a being, for example a severely cognitively and physically disabled human infant, would have a desire to stay alive under ideal conditions now, a realistic assessment of how the infant's future life will look like is needed. Ideal conditions cannot be taken to mean that the child would not be cognitively disabled, because then it would be in the child's interest to learn how to read, for example, which would be absurd. Clearly, it is the cognitive disability itself that contributes to the child's specific needs. If ideal condition are rather taken to mean that we should try to decide what would be best for the child from a well-meaning, paternalistic perspective, we are moving away from a desire-based theory towards a position that puts the subject's welfare into its focus. If this is correct, however, then it is not the existence of a non-satisfied desire *per se*, that leads us to say that someone is harmed by her premature death, but the fact that a being is deprived of potentially hedonistic or valuable experiences.

2.2 Loss of Future Goods

From what has been said so far, one could get the impression that death's harm is in fact all about the deprivation of future goods[12]. This is the position of so-called *life comparative accounts*. The more potentially good experiences death prevents from being enjoyed, the worse a particular death *ceteris paribus* is. In determining whether a specific death was harmful, the goodness of the (actual) shorter life is being compared with the goodness of the longer life[13] the individual would have had (see, e.g., Broome 2013). This is on par with a widely held concept of benefits and harms that Steven Luper summarizes as follows:

> An event E is in S's interests just in case E overall benefits (is good for) S, making S's life better than it would have been if E had not occurred, which E does just when its value for S is positive. An event E is against S's interests just in case E overall harms (is bad for) S, making S's life worse than it would have been if E had not occurred, which E does just when its value for S is negative. How much E benefits (harms) S depends on how much better (worse) S's life is in the actual world than it would have been if E had not occurred: the better (worse) S's life is, the more beneficial (harmful) E is. (Luper 2009, 6)

On the first glance, this account has great intuitive appeal, because it implies that death is *ceteris paribus* worse for a subject, the earlier in life it occurs and also, that death is generally worse for one individual at thirty than for another at eighty. On the second glance, however, this account proves to be rather unsatisfactory, because it also implies that the death of a one-day-old infant should be considered more harmful than the death of an older child. In fact, an abortion shortly after a subject has come into existence would preclude the maximum number of pleasures the victim could possibly have reached and thus be considered as worst case. Very few people will be prepared to accept that (see also McMahan 2002, 165–71). While it is generally true that death at a younger age is worse than death at an advanced age, our intuitions also seem

12 I will use the term *future goods* to include not only pleasures that directly add to a subject's hedonistic state but also experiences that do not always come along with positive emotions, but which the subject values nonetheless, for example, the good of gaining knowledge or of making a contribution to society. These sorts of valuable experiences seem to play a major role for our definition of a good life as human persons. Of course, it could be argued that these kind of values, too, are only valued in turn, because of their effects on our well-being or social status.

13 How to specify this hypothetical longer life against which the actual, shorter life must be compared, is a rather complex question. As it applies to McMahan's time-relative interest account as well, it will be discussed in detail in chapter 3.2.

to suggest a turning point somewhere during infancy or adolescence, at which the death-at-a-younger-age-is-worse thesis is reversed.

Jeff McMahan now argues that the reason for why the death of a 15-year-old is commonly taken to be worse than the death of a newborn infant is that even though the loss of future good that the newborn suffers is immense, her mental life is extremely limited, she has no desires or intentions with regard to her future and if she would have lived to become a person, she would not have remembered life as an infant. The 15-year-old, in contrast, has desires and plans for his future, he has expressed the wish to become a magician, has developed certain character traits and beliefs he will continue to hold as an adult, and so on. In the case of the newborn, in contrast, there is a sense in which the loss suffered from death almost seems to be detached from the specific subject: "It is almost as if the future [the newborn] loses might just as well have belonged to someone else" (McMahan 2002, 170).[14] It seems, therefore, that besides the loss of future good, multiple other aspects contribute to death's harmfulness, for example, personal sacrifices and investments for the future, unfulfilled life goals, desires and so on and so forth.

Some advocates of a life comparative account might still want to embrace its conclusions regarding abortion and would want to stick to this theory. There is another reason, however, for rejecting it. Life comparative accounts measure the badness of one particular death in terms of the effect it has on the overall value of the individual's life as a whole (McMahan 2002, 105). In comparing the loss of good of subject A_1's hypothetical longer life with subject A_2's actual shorter life, the life comparative account presupposes A_1 and A_2 to be one and the same individual. Otherwise there would be no way to determine exactly which future goods must be included in the calculation. Consequently, life comparative accounts imply that loss presupposes identity: one can only lose a future good and be harmed by its loss, if the present and the future person are one and the same. Life comparative accounts implicitly claim that in order to be *concerned* about a future event, the future self, who will experience it, must, as a basic requirement, be identical with one's present self, that is, the concerning self.

Although intuitionally one might take this to be almost self-evidently true, McMahan holds that Derek Parfit's (1987) notorious Division cases sufficiently show this intuition to be false, as they clash with other strong intuitions most of us share. Although person P_1 cannot become both P_2 and P_3 in Division, he is rationally justified in being concerned about both personalities (for a detailed discussion, see McMahan 2002, 40ff). This famously led Parfit to the view that

14 This statement will be further qualified and, to a certain extent, relativized, in chapter 3.4.

what *matters*, to lend Parfit's famous phrase, is not identity, but the underlying psychological relations binding a subject to his future. The stronger this connection, the stronger the subject's *rational egoistic concern*[15] about his future.

In practice, these psychological relations fall together with identity, however, in extreme circumstances like the Division cases, "it can [...] be rational and appropriate to be egoistically concerned about a person in the future even if that person will not be oneself" (McMahan 2002, 42). I will come back to this aspect, as it constitutes an important element of McMahan's theory, but for now it is enough to mention it as a further argument against life comparative accounts.

In conclusion, neither desire-based nor life comparative accounts completely capture the spirit of death's harmfulness. Desire-based accounts are either too cognitively demanding in that they preclude too many beings or they fall into hedonistic theories. Life comparative accounts cannot account for the intuitive belief that death does more harm to a child or a young adult than it does to a developed fetus or a newborn infant. Moreover, defining death's harm through the overall loss of future goods one suffers implies that personal identity is decisive here, an assumption standing on shaky grounds, according to McMahan.

On the other hand, both accounts emphasize certain important aspects a plausible ethics account of killing must include. This is exactly what Jeff McMahans *Time-Relative Interest Account* (TRIA) accomplishes. McMahan (1998) also refers to it as a weighted future goods account, because it evaluates a particular death's harm according to the loss of future goods the victim suffers, but additionally takes the psychological bond persisting between the subject at the time of her death and herself at the time the goods would have occurred into account.

15 As rational egoistic concern can hold in the absence of identity, according to psychological accounts, it would be more accurate to talk about "special, egoistic-like concern" (McMahan 2002, 42), as McMahan himself notes.

CHAPTER 3

The Time-relative Interest Account of the Badness of Death

3.1 Time-relative Interests versus Welfare Interests

In the foregoing chapter it was argued that if we want to save the beliefs that, on the one hand, animals, infants and mentally disabled human beings can be harmed by their deaths, but also, on the other hand, think that death does more harm to a child than it does to a newborn infant, then we are in need of an alternative account of the badness of death. Jeff McMahan now proposes that how much someone is harmed by his death is not only a question of how much future good his life would have contained, but also, how much he himself is psychologically connected to that future life he loses in the moment of dying. Based on these considerations, he formulates the following principle, which states that death's harm is primarily[16] a function of two elements[17]:

> The *Time-Relative Interest Account* of the badness of death evaluates death in terms of the effect that it has on the victim's time-relative interests rather than on the value of his life as a whole. It holds that the badness of death is proportional to the strength of the victim's time-relative interest in continuing to live. The strength of his time-relative interest in continuing to live is a function of both the net amount of good his life would contain if he were not to die, and the extent to which he would be bound to himself in the future, if he were not to die, by the prudential unity relations. (McMahan 2002, 105f)

16 Other factors additionally contribute to the badness of a specific death, according to McMahan, but many of these other factors are already indirectly included in these two, as will be clarified below.

17 It is silent, however, about the relative weights of these two factors which opens significant room for interpretation. If the prudential unity relations are of maximum strength, McMahan proposes to multiply the value of a future event it will have when it occurs by the factor 1, and accordingly multiply it with a fraction of 1, if the prudential unity relations are weaker (McMahan 2002, 80). Of course, this still leaves the room for interpretation wide open, as it is not clear at all how to deal with cases where, on the one hand, individual A's loss of good is very high and the prudential unity relations are very weak, and on the other hand, B's loss of good is moderate and the prudential unity relations very strong. McMahan regards this as advantageous, because a theory should mirror our uncertainty in cases like these. I agree.

 | DOI:10.30965/9783957437365_004

Prudential unity relations are those "relations that provide the basis for rational egoistic concern about the future" (McMahan 2002, 42) and will be discussed in full detail in chapter 3.3. Before coming to the two elements of the time-relative interest account (TRIA), however, a few more words need to be said on *time-relative interests* and their conceptual delineation from the two other kinds of interests, that have been differentiated above, that is, welfare interests and preference interests.

What exactly interests are and what determines their strength has been controversially discussed and this thesis is not the place to give a satisfactory review. The following remarks will be necessarily scarce, but hopefully clarify the concept of time-relative interests nonetheless. Interests, commonly understood, are components of an individual's well-being. If something is in the interest of an individual, then its satisfaction will make life better for this individual (see also Cochrane 2007). This understanding of an interest thus excludes bad or uninformed desires, such as the desire to take drugs or Romeo's desire to join Julia in death. The satisfaction of a desire can be in an individual's welfare interest, but not necessarily.

It can be notoriously difficult to determine whether the satisfaction of a certain desire really is in the interest of an individual or whether it would be actually better for her if her desire remained unsatisfied. A novelist, for instance, may agonize over her work, even destroying parts of it only to rewrite them later or even stop writing for several years. Even if the author herself held that the suffering had been worth the result in the end, it might be that she is just wrong from a welfare perspective, in that she would have led a happier life if she had just become a gardener rather than a novelist. However, we also ascribe certain experiences with value that do not necessarily make us happier. Well-being or welfare, thus, can be understood to include more than just an individual's level of happiness and should allow for more things to be in one's interest. This aspect will be further discussed in chapter 3.3.2, focusing on McMahan's concept of psychological unity. To conclude, even though it can be unclear whether something should be considered as contributing to an individual's well-being or not, it will nonetheless be assumed that interests are constituents of an individual's well-being.

To count as a loss or harm the subject suffers, it is enough to know that the subject would have enjoyed that particular good the moment it would have occurred or that it would have contributed to her overall well-being. To have a preference interest – for example, a desire – for a particular good at the time of dying is not a basic requirement in order to be harmed by the loss of that good. Life comparative accounts determine death's harmfulness according to the loss of future good the individual suffers. Because of their counterintuitive results and their disputable presupposition that identity is a necessary condition in order to be harmed by death, however, McMahan introduces a third concept of interests.

> One's interests are concerned with what would be better or worse for oneself as a temporally extended being; they reflect what would be better or worse for one's life as a whole. If identity were the basis of rational egoistic concern, there could be no divergence between one's interests and one's time-relative interests. For the prudential unity relation – identity – would not be a matter of degree and would always be present whenever there was any basis for egoistic concern. In short, if identity were the basis of egoistic concern, the concept of a time-relative interest would be otiose. But if identity is not the basis of egoistic concern and the prudential unity relations can hold to varying degrees, we need a concept that is temporally relativized – one that expresses the idea that the degree of rationally warranted egoistic concern over some possible event may vary over time depending on the strength of the prudential unity relations. (McMahan 2002, 80)

Time-relative interests, in short, "are what one has egoistic reason to care about *now*" (McMahan 2002, 80). Contrary to welfare interests, they are related to the being's current psychological status. Contrary to preference interests, the individual is not required to have a desire for this future good. Time-relative interests, thus, also include welfare interests for which the individual at the time of her death does not have a desire for. This allows for something to be in the time-relative interest of an individual, for instance, a newborn baby or an animal, even if the individual herself is unaware of this future good.

Elizabeth Harman (2011) criticizes McMahan for not being able to account for our intuitions in certain surgery cases. I agree with McMahan (2015) that her examples in fact do not pose real objections to the TRIA, because they rest on the false assumption that the TRIA only considers an individual's present interests. As has been clarified above, time-relative interests also accommodate certain future interests the individual does not yet hold. What an individual has egoistic reason to care about now, that is, a being's time-relative interests, is different from the individual's present preferences.[18]

---

18 McMahan acknowledges, however, that Harman rightly points out a major challenge to the TRIA and as a result comes up with several other surgery cases. Without wanting to go into great detail here, the basic problem these cases reveal is that the TRIA seems to put too much weight on the good or bad lying in the near future of an animal and thereby disregarding the overall goodness or badness of the animal's future life. This can lead to the unintuitive result that, on the one hand, an animal, whose overall prospect in future life is good, should nevertheless be allowed to die according to the TRIA, because she would otherwise face a few days of moderate suffering in the near future, but that, on the other hand, an animal with an overall negative prospect in future life should be

3.2 Determining the Loss of Future Good

3.2.1 *The Problem of Specifying the Antecedent*

According to McMahan, the badness of a death is to be determined, among others, according to the net amount of good the individual's life would have contained, which raises the question how this imaginary longer life should plausibly be assessed. Besides the epistemological problem that, naturally, we are unable to predict a hypothetical future with accuracy[19], this poses a metaphysical problem as well to which McMahan refers to as the *problem of specifying the antecedent*. We are to imagine what would have happened if the individual had not died. However, there are indefinite possible worlds, where this would be true. Are we to accept, then, that one death can be good with regard to x (painfully dying from cancer two years later), but bad with regard to y (happily living on for twenty more years)? This is extremely unsatisfactory; we want a "single, general, context-independent evaluation which tells us whether, all things considered, [a] death was good or bad, period" (McMahan 1988, 43).[20]

McMahan proposes a useful, though possibly not exhaustive, set of criteria, as he himself acknowledges (for a more thorough analysis, see McMahan 1988 and McMahan 2002, 107–145). Taken together, the hypothetical longer life serving as a comparison should (1) "completely exclude the cause of death and its effects", (2) represent the best future possible to the one who died, (3) "preserve

saved, because otherwise her moderately strong interest in the pleasurable experiences lying in her near future would be thwarted. McMahan proposes several solutions how to circumvent these unintuitive conclusions within a TRIA. These solutions are dependent on a series of other controversial assumptions, such as the (*Weak*) *Asymmetry* view, that states that while the anticipation that an individual would have a life that would be bad for her gives us a reason not to bring her into existence or allow her to live, the anticipation that an individual would lead a life that would be good for her does not give us a reason, or only a very weak one, to bring the said being into existence or allow her to exist (McMahan 2015, 76). The thorough analysis, that would be necessary to investigate McMahan's eight surgery cases and the extreme complex and premise-rich solutions McMahan proposes in order to weaken them lies well beyond the scope of this work. I therefore refer interested readers to McMahan (2015).

19 This problem cannot be solved. This does not imply, however, that evaluations how well or bad a certain life will fare, based on knowledge of what typically influences the course of life, can or should not be made.

20 Even though McMahan later, in his *Ethics of Killing*, is more tentative in that he acknowledges that in some instances death might rightly be described as more or less harmful with regard to x or y, he considers these cases to be rare. In most cases, a single evaluation of the badness of a specific death is thus possible.

[one's] earlier life as it actually was" and (4) represent "a realistic possibility, though not necessarily a practical possibility in the circumstances" (McMahan 2002, 116f).

With regard to the first criterion, McMahan proposes to specify the world that is closest to the actual world where the *transitive cause of death* is absent. The main challenge, therefore, is to determine the transitive cause of death which is "the set of all events that form part of the chain of causes leading to C", where "C is the immediate or proximate cause" (McMahan 1988, 47) of death.[21] The first and the third criteria stand in a dynamic conflict in cases where death was not caused by such a random and unrelated event as an accident, for instance, but where "the chain of causes" includes many events. This can lead to very different interpretations of what caused death, as can be exemplified by help of McMahan's example of the cavalry officer, who is shot and killed by the Russian soldier Ivan. Boris, another soldier, simultaneously fired a bullet on the cavalry officer, so he would have been killed anyway seconds later. In his wide interpretation (McMahan 1988, 45), McMahan defines war as the transitive cause of the cavalry's death, in his narrow interpretation (McMahan 2002, 118), he merely imagines that the cavalry was out of Ivan's and Boris' reach on a different area of the battlefield.

In spite of this, the second and the fourth criteria make sure that these differences decrease again in most cases. Consider the killing of a rearing pig. The immediate cause of the rearing pig's bleeding to death is the transection of the carotid artery, let us assume. Analogously to McMahan's wide interpretation in the cavalry officer example, one might argue that the pig's transitive cause of death lies in the simple fact that he has been raised with the sole purpose to be killed for food one day. If that transitive cause is absent, the pig's real, short life should not be compared to a longer life with his inevitable slaughter at its end, but to a life as a pet, for instance.[22]

---

21 In the *Ethics of Killing*, McMahan does not use the term transitive cause of death anymore. While discussing cases of overdetermination, he acknowledges, however, that the cause of death also includes "parts of the same causal sequence" (McMahan 2002, 119).

22 Due to his genetic conditions as a rearing pig, he would not live as long as an average pet pig. The closest possible world would possibly be one where the pig died, after a happy pet life, at the average age of rearing pigs. Theoretically, it is possible to argue that the closest possible world would be one where he would not be a rearing pig – including the genetic conditions coming along with this –, but a wild boar. See also McMahan's discussion of Genette who dies from a genetic defect (McMahan 1988, 53). I take this last option to be much more controversial, however, and also in conflict with McMahan's later stated requirement to preserve the individual's earlier life as much as possible (see above).

It is equally possible, nonetheless, merely to imagine that the pig somehow escapes the slaughterhouse, analogously to the cavalry officer escaping the battlefield. As we are to imagine the best future possible to the animal and this best future must be "a realistic [...], though not necessarily a practical possibility in the circumstances" (McMahan 2002, 117), we should probably assume that the pig would not be caught, or that he would be found by a sympathetic person. Either way, the pig is clearly being deprived of something from his death in the slaughterhouse. Most importantly, his slaughter cannot be considered as an act of mercy.

That the killing of rearing animals should be considered as a beneficial act, has been repeatedly put forward, more recently, for instance, by food journalist Michael Pollan (2007). This view emphasizes that the transitive cause of the pig's death – people want to eat pork – in this case is identical with the cause of the pig's existence. If the transitive cause of his death is absent, the pig would not exist at all. The pig's actual short life must therefore be compared to his hypothetical non-existence. Non-existence, the argument goes on, is worse for the animal himself than a short, overall happy life. Therefore, we actually benefit the pig by rearing and killing him. Henry Salt called this argument rather disparagingly the *logic of the larder* (Singer 2011, 107). Even if one assumes, for the sake of the argument, the rather dubious claim that the rearing pig would enjoy an overall happy life until his death, the larder's assumption that non-existence can benefit or harm an individual must be rejected. Non-existing individuals do not have zero welfare, they have no welfare at all (see Višak 2015 for a similar argument).

Applied to animals, the transitive cause argument therefore reveals the fallibility of a widespread argument adduced against the harmfulness of rearing animals' deaths. The actual, shorter life is compared to a hypothetical, longer life that is statistically the most probable. Thus, the death of a rearing pig at some specific point in time t_1 is not considered to be very harmful, because the pig would have been killed regardless. It would have been spared some more weeks, let us say, but if these weeks would have been all in all painful and distressing, the argument goes on, it might even be better to kill the animal rather sooner than later.

This kind of argument, however, is fallible. The net amount of future good, that is being taken away from the animal, must not be determined in comparison to a hypothetical longer, statistically-speaking most probable life, but to the *best* (realistic) future the animal could have. To sum up, a calculation of the net amount of future good a life would contain if the individual were not to die, must be made by reference to the closest hypothetical world, in which the transitive cause of death was absent. It should be kept in mind, however, that this might still lead to controversies from time to time.

3.2.2 *The Problem of Cross-species Comparisons*

> A dog can never tell you what she knows from the
> smells of the world, but you know, watching her,
> that you know
> almost nothing. (From Mary Oliver, *Her Grave*)[23]

Another question has yet to be answered, that is, how the loss through death is to be calculated exactly, and, importantly, how one's individual loss is to be compared with another one's. These are no trivial questions, as we will see in the following paragraphs, and raise some serious objections to McMahan's TRIA.

McMahan argues that the loss of future good the individual suffers from a death is "a function of […] the net amount of good his life would contain if he were not to die" (McMahan 2002, 106). The net amount of good of one's future life is calculated by subtracting the total amount of bad from the total amount of good this future life will plausibly contain. This implies that, in theory, the net amount of future good can be negative. Such a miserable being had a negative time-relative interest in continuing to live, then, regardless of the few positive goods still lying in prospect.[24] Here and hereafter, it will be assumed that the net amount of good an individual loses from death is positive.

In comparing the net amounts of future good of two different individuals, the TRIA calculates these goods in absolute, rather than in relative terms. The amount is not determined relative to what the individual in question can achieve, due to her natural (cognitive) dispositions, for example, but according to what is maximally possible at all. There is only one scale of lost good, so to speak, for all kinds of beings having an interest in continuing to live. Yet, in determining this loss, it is not only the number of future goods, that is decisive, but also their quality. This becomes particularly clear in one of McMahan's earlier formulations of the TRIA:

> The badness of any particular death will be determined by [i] the quantity and quality of the future goods (minus future evils) that the individual's life would otherwise have contained, though [ii] the badness of the loss of each future good through death must be weighted for the degree to which the individual at the time of death would have been psychologically related to himself at the time the

23 The above quotation is an excerpt from Mary Oliver's poem "Her Grave". It has been published in a selection of poems (see Oliver 1992, 14).

24 Such a miserable individual still suffered a misfortune in dying, however, because McMahan acknowledges that being in a position of having nothing more to lose can constitute a misfortune, that the TRIA does not capture (for a detailed assessment of three different kinds of misfortune, see McMahan 2002, 107–140).

> good would have occurred in his life had he lived rather than died. (McMahan 1992, 22)

Clearly, in most cases, an animal has "relatively little good in prospect" (McMahan 2002, 199) *compared to* what an average human person has in prospect.[25] To determine the loss an animal suffers through death by comparing it to the loss a person suffers, however, raises questions of justification. Even if one accepts that, generally, most animals have a weaker time-relative interest in continuing to live than most human adults, one could argue that this does not capture the essence of death's harmfulness, because it represents a cognitively chauvinistic view. If this turned out to be true, the TRIA would be incapable of determining death's harmfulness for two individuals of different species, which do not, and never will, share a similar psychology, for example a dog and a human person. The objection, therefore, is serious for my research question.

McMahan quite confidently states that a person's and a dog's losses from death are comparable. According to him, it is even "obvious" that "the life an animal loses in dying is almost always inferior in quality and quantity to the life that a person is deprived of by death" (McMahan 2015, 70). According to the TRIA, therefore, the passengers in Regan's lifeboat should sacrifice the dog, because, leaving the question of psychological unity aside here, each one of the prolonged human lives will contain a higher amount of good than the dog's life would or could.

Clearly, human persons have access to many very elaborated pleasures. Even if we grant a dog to greatly value certain experiences like eating, running, dozing in the garden, chasing the neighbor's cat, and so on, he will never be able to take (aesthetic) pleasure in music, art or literature, writing a book or baking a complicated chocolate pie, due to his constitution. In the words of McMahan:

> Because of their limited cognitive and emotional capacities, most animals lack the capacity for many of the forms of experience and action that give the lives of persons their special richness and meaning, and without which our lives would be greatly impoverished. Animals are incapable of deep personal relations based on mutual understanding; they lack both imagination and an aesthetic sense and hence are unable to experience works of art, literature, or music or to appreciate

25 Besides the quantity and the quality of future goods, McMahan mentions additional reasons why the life of an animal normally contains much less good than a person's (McMahan 2002, 197f). Most of these strongly correlate to the extent of how much someone is psychologically bound to oneself in the future. I will come back to them in chapter 3.3.2, when delving deeper into the notion of psychological unity.

> the aesthetic dimensions of the natural world; they are incapable of engaging in complex and skilled activities or achieving difficult goals or ambitions – for example, through artistic creation or scientific discovery; and so on. (McMahan 2002, 195)

In another passage, McMahan – a great dog-friend himself, one may add – humorously states that "[h]owever long I set my dog in front of my print of Goya's *Devota Profesión,* he will never derive the least spark of aesthetic gratification from it" (ibid. 149). This is obviously correct. But does it also matter in a morally relevant way? How do we get from the existence of allegedly more and less advanced pleasures to the conclusion that these qualitative differences are also morally relevant? This is not trivially true.

One possibility would be to argue from an objective perspective (also called *objective list accounts*), stating that certain capabilities are just good in themselves. This is indeed McMahan's position, mirroring Mill's famous conclusion that the life of a human person is always to be preferred to an animal's life, even if the human life was miserable and the animal's life extremely fortunate. If perfectionist views are to be rejected, however, because they invite concerns about being dogmatic (why should the existence of beings with a capability x *per se* be preferred to a universe without these beings?), such a claim requires further justification. According to Mill, his statement is compatible with his greatest happiness principle, because certain intellectual pleasures have a higher value than so-called lower pleasures.

> If I am asked, what I mean by difference of quality in pleasures, or what makes one pleasure more valuable than another, merely as a pleasure, except its being greater in amount, there is but one possible answer. Of two pleasures, if there be one to which all or almost all who have experience of both give a decided preference, irrespective of any feeling of moral obligation to prefer it, that is the more desirable pleasure. If one of the two is, by those who are competently acquainted with both, placed so far above the other that they prefer it, even though knowing it to be attended with a greater amount of discontent, and would not resign it for any quantity of the other pleasure which their nature is capable of, we are justified in ascribing to the preferred enjoyment a superiority in quality, so far outweighing quantity as to render it, in comparison, of small account. (Mill 1863/2008, 10)

It should be self-evident, however, that you and me will never be able to properly take a pig's or a dog's perspective and judge the pleasures of wallowing in a mud puddle. Mill's judge of someone who is "competently acquainted with both" must be a mystical, godlike character – Singer gives the example of Shakespeare's Puck – but we, as human beings, could never slip into the consciousness of a pig and a dog and afterwards compare these two existences to ours from a neutral point of view. We might be able to prefer life of a dog over life over a polar bear from our point of view, including our needs for social relationships, our climatic preferences, our specific cognitive background and the values we are acquainted with, but this is far from what Mill has in mind. Ben Bradley comes to a similar conclusion:

> It is a very difficult thing to compare the values of lives of members of different species. I don't know how to tell whether my dog is having a better life than mine or a worse life; I prefer my own life, but that is compatible with it being the case that his life is better for him than mine is for me. In thinking about this question, it is very difficult to be objective – almost all of us want to say that lives like ours are more valuable than the lives of very different kinds of things, at least in part because this justifies us giving preference to beings like us. (Bradley 2015, 59)

In the face of these difficulties, I believe that the attempt to justify the preference of certain higher pleasures from a hypothetical quasi-neutral position must fail. Another possibility might be to rather take a more hedonistic point of view; some experiences, it might be argued, simply contribute more to one's level of well-being than others. Just like he can value the pleasure of taking his dog for a walk more than the pleasure of enjoying a cup of tea, McMahan's pleasure in looking at Goya's painting simply outcompetes *any* of the pleasures his dog is capable of.

However, even if, for the sake of the argument, it is accepted that complex cognitive capacities really amount to more happiness – a highly controversial claim (see, e.g., DeGrazia 1996, 238–240 and also McMahan 2002, 195) – one might still object that a determination of the dog's loss in comparison to the person's loss is quite unfair and implausible. How can the dog be said to have an inferior life, simply on the grounds that he does not have the natural dispositions to take pleasure in visiting a museum? The dog himself does not care, so who else should? Alasdair Cochrane argues, in a similar fashion:

> [W]hen we talk about the "net amount of good" in a life, this raises the question, good for whom? In McMahan's account, when determining the strength of an interest the value of the good *for the individual whose life it is* does not matter, all that matters is that there is "more net good". But this takes us away from the notion of interests and well-being as prudential values. If [...] the interests of an individual relate to how well life goes for the individual whose life it is, then an interest must only be stronger if it relates to a good that is of more value *for that individual.* (Cochrane 2007, 311)

To avoid these problematic implications, the TRIA could be modified to the effect that it does no longer calculate two individual's net amounts of future good in absolute terms and compare these absolute numbers to one another, but relative to what would be maximally possible, in the first place, for the individual in question (for similar approaches, see Višak 2017 and Cochrane 2007, 2012). This modified time-relative interest account determines the losses that the person and the dog suffer from their own perspectives, which are then set in comparison to each other. For example, if both McMahan and McMahan's dog were considered to reach one hundred percent on their individual scales of welfare in the future, they should be considered to lose equally much future good from their deaths, even under the assumption that McMahan can enjoy many more things than his dog can.

It might even be argued that a pet dog comes much closer to the highest attainable amount of well-being he has the disposition to achieve than a person does. Persons can enjoy an enormous amount of potentially valuable experiences. In a way, persons are more *demanding* in what they expect from life than a dog probably is. Even a very well-off person, therefore, would arguably still fall short of her potentials. Thus, a dog could be argued to actually lose *more* from death than a person. Similarly, the amount of good a frog loses from his death, may be considered to be higher than the amount of good the dog loses.

This, on the first glance rather unintuitive, statement may become more plausible in combination with the TRIA's second element, because the amount of good that death thwarts would still have to be weighted for the degree of psychological relations, binding an individual to his future self. Thus, even if it was acknowledged that the loss of future good should in fact be considered greater for the dog than for McMahan, his much stronger prudential unity relations would still make McMahan's interest in continuing to live greater than his dog's.

Unfortunately, a thus modified time-relative interest account raises many questions and problematic implications on its own, which would certainly be worth investigating in full detail at another time. The focus of this thesis, however, is on future-oriented cognition of animals, so I will turn my attention now to the TRIA's second element, namely the prudential unity relations. Nonetheless, I acknowledge that the first element of McMahan's TRIA raises some serious problems, because it determines an individual's loss in terms of the net amount of good, rather than according to the value these goods have for the individual.

3.3 The Prudential Unity Relations

3.3.1 *Psychological Connectedness and Continuity*

The time-relative interest account states that "[t]he strength of [an individual's] time-relative interest in continuing to live is a function of both the net amount of good his life would contain if he were not to die, and the extent to which he would be bound to himself in the future, if he were not to die, by the prudential unity relations" (McMahan 2002, 106). The prudential unity relations are those relations, that provide the basis for rational egoistic concern. Their strength varies with the degree of *psychological unity* between oneself now and oneself in the future, which, in turn, encompasses both *psychological connectedness* and *psychological continuity*. Here, McMahan largely adopts Derek Parfit's terminology. The following paragraphs will provide definitions of these three concepts and explain how they are related to one another.

How strongly someone is related (or bound) to oneself in the future is a function of the degrees of psychological connectedness and psychological continuity. A person can be weakly or strongly psychologically connected to her future self, depending on the number of *direct psychological connections* between herself now and the future. Parfit (as well as McMahan) defines these connections ostensively, that is, via examples:

> Besides direct [experience-]memories, there are several other kinds of direct psychological connection. One such connection is that which holds between an intention and the later act in which this intention is carried out. Other such direct connections are those which hold when a belief, or a desire, or any other psychological feature, continues to be had. (Parfit 1987, 205)[26]

Suppose that A_n represents a temporally distinct "self" of individual A at time t_n. Now, when between A_1 at time t_1 and A_2 at t_2, there are at least half the number of direct psychological connections that there are between "nearly every actual person" (ibid. 206) on two consecutive days, then this establishes a *strong psychological connectedness* between A_1 and A_2. If there are less than half the number of direct psychological connections, there is only *weak psychological connectedness*.[27]

---

26 The continuity of certain character traits is another example (Parfit 1987, 301).

27 Obviously, Parfit's transition from weak to strong connectedness is rather arbitrarily defined. Not only is it unclear what a normal person, i.e., "nearly every actual person", in Parfit's words, is, or how many direct psychological connections occur between a normal person on two consecutive days, it is also opaque why fifty percent of connections should be normatively relevant. Finally, it is very artificial to assume that we can count psychological connections at all.

Psychological connectedness is intransitive, which means that from the fact that A_1 and A_2 and A_2 and A_3 are psychologically connected, it does not *per se* follow that A_1 and A_3 are connected as well. However, even if A_1 and A_3 are not directly psychologically connected, they can be *psychologically continuous*, if direct psychological connections occur between A_1 and A_2 and A_2 and A_3.

Here, McMahan discerns from Parfit, who defines psychological continuity as "the holding of overlapping chains of *strong* connectedness" (ibid.), which means that psychological continuity only occurs in an all-or-nothing manner. Contrary to Parfit, McMahan defines psychological continuity as a matter of degree, depending on the strength of their underlying chains of (weak or strong) connectedness. If the direct psychological connections that bind A_1 and A_2, and A_2 and A_3 together are strong, the relation of psychological continuity between A_1 and A_3 is strong, too. If they are only weak, A_1 and A_3 are only weakly psychologically continuous, in turn. In short, for McMahan – but not for Parfit – psychological connectedness *and* psychological continuity occur in degrees.

In defining psychological continuity in gradual terms, as McMahan does, the question arises how to deal with overlapping chains containing varying degrees of connectedness. McMahan seems to hold that if between A_1 and A_n there is at least one instance of only weak connectedness, A_1 and A_n are only weakly continuous as well, because he also assumes that an adult person is only weakly continuous with herself as a baby. Of course, this also implies that the same individual, suffering an extreme psychological trauma by the age of forty, might only be weakly psychologically continuous with her pre-traumatic self as well.

These two aspects, psychological connectedness and psychological continuity, determine how much someone is related to her future self and to what extent she and her future self constitute a psychological unity. Even though psychological unity encompasses psychological connectedness and continuity, it is more than just the sum of these two relations, because it varies not only with the degree of its underlying notions, but also with the *kind* of psychological connection.

Having given a first overview of the nature of the psychological relations that bind an individual to its future self, let us take a look at the comparative implications the TRIA has. Consider again the two deaths of a 15-year-old child and a newborn infant. Remember, the comparative life account was rejected by McMahan, because it had not been able to represent most people's intuition that death would be worse for the child (even though it was also acknowledged that intuitions may considerably diverge). The TRIA, can accommodate this intuition: even if it is assumed that the loss of future goods would be much higher in the case of the baby (because he has his whole life ahead of him), the TRIA states that death would do a much greater harm to the student, because she is much stronger psychologically related to her future self than the baby is. Between the baby and himself as a future person there are no direct psychological connections such as memories, intentions and desires at all. Moreover, he is only very weakly psychologically continuous with his future self, because at least in the first year of life of a human being, there are only few of those direct psychological connections from day to day.

The youth, in contrast to the baby, is quite strongly related to herself at the times the goods that would have laid ahead of her would have occurred: she is very strongly continuous with her future, and there are many direct psychological connections between herself now and her future self like the formation, the persistence and the satisfaction of the desire to become a physicist. The TRIA indirectly acknowledges, therefore, that desires play a significant role in the harmfulness of a particular death and agrees with Bernard Williams insofar that the more categorical desires death prevents from being satisfied, the worse the loss of these goods becomes. It disagrees with the desire-based account, however, that the thwarting of desires lies at the core of death's badness.

How does the death of an animal now relate to the death of a person, on the one hand, and the death of an infant, on the other hand, in this approach? Although psychological connectedness and continuity are defined by reference to normal human persons only, McMahan applies it to all other kinds of sentient beings as well. That means that an animal's degree of psychological connectedness, for example, is determined in direct comparison to a normal person. Assuming that the number of direct psychological connections occurring between t_1 and t_2 is less than half the number there is, over the same period, in the life of the person, the animal is said to be only weakly psychologically connected.

Based on its two elements – the (net) amount of future good and the extent to which someone is bound to oneself in the future by the prudential unity relations – the TRIA gives rise to a scale. At its lowest end is non-conception (extremely high amount of lost future goods, but no subject)[28], at its highest end is the death of a (human) person (high amount of lost future goods, very strong prudential unity relations). Somewhere in-between are the deaths of

28 Of course, where there is no subject, there is no harm that can be done to her, so non-conception cannot actually be said to represent the lowest time-relative interest in continuing to live. The death of a human being shortly after having come into existence should actually be a better example, consequently.

an infant (or developed fetus) and animals. As in the case of non-conception, the infant loses an extremely high amount of future good. Yet, first, there are no direct psychological connections between the baby and his future self as an adult, and second, he is only weakly psychologically continuous. Even assumed that an animal might be slightly more continuous with her future self than an infant, McMahan believes, as has been seen, that she has "relatively little good in prospect" (McMahan 2002, 199). He comes to the conclusion, therefore, that an infant normally has a stronger time-relative interest in continuing to live than an animal does. In summary, the TRIA states that, *ceteris paribus*, the death of a person is worse than the death of an infant than the death of an animal.

Based on the criticism of the foregoing chapter, a modified TRIA arrived at a slightly different assessment than McMahan in that the death of an animal should be considered to be worse than the death of an infant, given McMahan's own estimation that animals are, or at least, can be, in fact stronger psychologically continuous with their future selves than infants. Moreover, McMahan acknowledges that some animals might be persons or near-persons, that is, in his definition, beings that could be described as (rudimentarily) autonomous (see also chapter 6.3.3). Those animals would consequently be located at the highest end of the scale on a modified time-relative interest account.

3.3.2 *Psychological Unity*

3.3.2.1 The Relation between Psychological Continuity and Unity

The concept of psychological unity remains relatively vague in *The Ethics of Killing*. The following paragraphs should thus be read as a tentative interpretation of McMahan's use of this concept. Psychological unity incorporates both psychological connectedness and continuity. In McMahan's account, all three concepts are matters of degree. It has become evident that the degree of psychological continuity directly arises from the strength of psychological connectedness. The relation between psychological connectedness and continuity, on the one hand, and psychological unity, on the other hand, is more complex. A strong psychological continuity normally, but not always, results in a strong psychological unity. *Vice versa*, a strong degree of psychological unity does not necessarily involve a strong psychological continuity. Below, two examples for these exceptional cases will be given.

Psychological unity is perhaps best understood in comparison to psychological continuity. Remember, both psychological continuity and unity are diachronic, they extend over a certain period of time. Psychological continuity consists of overlapping chains of connectedness. Therefore, the degree of psychological continuity directly depends on the strength of its underlying chains of connectedness. Overlapping chains of strong connectedness between t_1 and t_2 result in strong psychological continuity over the same period of time.

As strong connectedness, however, is defined only through the sheer *number* of direct psychological connections involved, the *nature* of these direct connections is irrelevant for the ascription of strong or weak psychological continuity. As long as more than half of the connections are being transferred from day to day, it does not matter how many of the connections at the start of the chain are identical to those at the end of the chain, whether different psychological connections semantically refer to one another or if they have nothing in common at all. This is due to the transitive nature of psychological continuity. One might say that psychological continuity is mainly about structure, in short, but not about contents.

This is different with psychological unity, which takes into account that mental states, separated over a significant amount of time, can be (and often are) semantically connected to one another, for instance, in case of desires, memories or plans. With regard to the strength of psychological unity, these connections have a higher significance than other connections, as will become evident in subsequent chapters.

Psychological unity is closely related to (and may sometimes be hard to distinguish from) another type of diachronic unity, which, in McMahan's theory, typically additionally affects the extent to which death is bad for an individual, namely, narrative unity. Narrative unity exists when the single parts of an individual's life can be brought into a "meaningful whole" (McMahan 2002, 175). Narrative unity, thus, lays even more emphasis on the content of psychological connections than psychological unity. A narrative needs a *theme* or a *label*, under which all the different desires, plans, projects, imaginations and so on can be integrated. Maybe the easiest example for such themes are long-term goals or plans, because they are so common for us as human persons.

Think about a 15-year-old girl's wish to become a famous internist. To accomplish this, she must receive a doctoral degree in medical science. In order to successfully apply for university, she needs a very good school certificate and concludes that she must study very hard from now on to reach her goal. Because this goal involves many steps in-between and stretches very far into the future, it may also be called a life goal. In this example, the goal of becoming an internist strongly psychologically unifies quite a long period in the life of the girl and also provides it with a narrative unity.

Yet, even if the girl changes her mind by the age of 22 and pursues a completely different life goal from that moment on – whale watching in Norway, say –, it does not necessarily compromise her overall strength of psychological unity, nor her narrative unity.[29] Let me clarify this further. Extreme volatility, understood as the rapid and repeated replacement of life aims, values and desires, in fact compromises the psychological unity of a person in McMahan's sense.[30] Volatility is often associated with shallowness or psychological unstableness. We tend to think that a volatile nature has not *found herself* yet and that her degree of psychological unity may not yet be as strong as it will hopefully become later. Changes in the personality or the character of a person, however, do not *necessarily* compromise psychological unity. The girl from the example above formed the desire to become an internist by the age of 15. Although it can be the case that she holds on to this wish for her whole life time, we would rather expect that during adolescence, old life aims will be replaced by new ones. Rather than regarding these radical changes as diminishing psychological unity, however, these are normally viewed as instances of personal growth, and are commonly valued as such.

Assume that the girl comes to the conclusion, after having spent three years on a whale watching boat, that she wants to see as much of the world as possible and that every year or the other she will start over new. Even such an ever-changing life might involve a great degree of psychological unity. The difference between this life and a life of volatility, first and foremost, is, that the latter has not been chosen deliberately. Volatility is defined by its absence of an over-arching desire, wish or project that is an expression of the person's will. Similarly, a Buddhist, aiming for a mental existence that is restricted to the very moment, can be strongly psychologically unified because of her higher-order desire to live her mental life as much in the present as possible (McMahan 2002, 82).

Consequently, McMahan gives these kind of higher-order desires that determine what kind of person one wants to be or what goal one tries to achieve in life, a very high significance for the question how strongly someone is psychologically unified. The ability to form higher-order thoughts is very closely related to autonomy, if autonomy is defined here with Dworkin as "second-order

29 Marc Slors, among others, criticizes that psychological continuity theories cannot account for rather extreme changes in the personality of a subject, due to conscious deliberation or gradual evolution (Slors 1998, 67f). Although McMahan, to my knowledge, does not directly respond to Slors, his concept of psychological and narrative unities can be read as a response to these and similar concerns.

30 Volatility is not to be confused with what has also been described as multi-talent or a scanner personalty. Scanners are interested in more topics than most other people are and often appear rather unsettled to their environment, because they may change their work more often than others, develop new hobbies and interests etc.

capacity to reflect critically upon one's first-order preferences and desires, and the ability either to identify with these or to change them in light of higher-order preferences and values" (Dworkin 1988, 108). Consequently, autonomy plays a major role in McMahan's account, as will become evident later.

To summarize, psychological unity remains complex and unfortunately relatively vague in McMahan's account. In his terminology it is diachronic and can refer to a shorter or longer period in the life of one individual or even refer to her whole life. It emphasizes the significance of those desires and other mental states that semantically refer to one another, and is closely related to narrative unity. Although psychological unity encompasses psychological continuity, the degree of psychological continuity does not always reveal the extent of psychological unity, at least in case of persons. A strongly psychologically continuous person can be extremely volatile, impoverishing the degree of psychological unity with regard to the person's whole life. *Vice versa*, a radical Buddhist, spending the majority of her mental life in the here and now, striving towards an ideal of atemporality, and who is psychologically continuous with herself to a much weaker degree than persons typically are, can form a strong psychological unity nonetheless. The reason for this lies, first and foremost, in the significance of higher-order desires for psychological unity in the lives of persons. In those individuals incapable of second-order thought, the degree of psychological unity is closely tied to the degree of psychological continuity. As will become evident in the next chapter, however, some psychological connections have a greater unifying effect than others.

3.3.2.2 Two Types of Psychological Connections

The last chapter clarified the phenomenon of psychological unity itself. This chapter will provide a closer look on what *constitutes* psychological unity. Psychological unity, once again, involves psychological connectedness and continuity. As the strength of the former, given a particular period of time, directly depends on the number of direct psychological connections involved, and psychological continuity, in turn, depends on the strength of connectedness, direct psychological connections form the building blocks of psychological unity. As has been seen, however, the relation between psychological connectedness and continuity, on the one hand, and psychological unity, on the other, is quite complex. It will be argued that this is mainly due to some direct psychological connections being more psychologically unifying than others.

There are many different instances of direct psychological connections, as, to mention just a few, the relation between an earlier and a later manifestation of a belief, the relation between the formation of a desire and the experience of the fulfillment of this desire, an experience of an event and the later act of remembering it, and so on. In determining the strength of psychological connectedness or continuity, this qualitative variety does not matter.[31] Strong connectedness, as defined by Parfit, holds whenever there are "over any day, *at least half* the number of direct connections that hold, over every day, in the lives of nearly every actual person" (Parfit 1987, 206). If there are less than half the number of direct connections, psychological connectedness holds to a weaker degree. Psychological continuity, according to McMahan (and in contrast to Parfit), is a matter of degree as well. It consists in the holding of overlapping chains of connectedness. If the kind of connectedness involved is strong, there is strong psychological continuity. If psychological connectedness holds to a weak degree, the psychological continuity is weak, too. Consequently, the degrees of psychological connectedness and continuity depend on the quantity of direct psychological connections involved. The overlapping chains are determined by quantity, one might say, but not quality.

This is very different in case of psychological unity. If only quantity mattered here as well, this allowed for compensation in (hypothetical) individuals possessing some direct psychological connections, but not others. Someone could have the ability to sustain beliefs through time, for example, but not form any intentions or desires. If the number of the sustained beliefs remained high enough over time, the individual could still form a strong psychological unity, which seems bizarre.

Parfit developed his notion of psychological continuity in the context of the personal identity debate. His aim was to show that identity does not matter for questions of survival.[32] Thus, he was mainly interested in human persons. The fact that direct psychological connections vastly differ in complexity, both in comparison to one another and with regard to their properties and characteristics in two different species, did not have to concern him. In the context of the personal identity debate, Parfit's method of simply quantifying the direct psychological connections to determine if something is a case of strong, weak or absence of psychological continuity, is sensible.

31 This has been criticized by many authors, among them, more recently, Schechtman (1994), and Slors (1998).

32 In a nutshell, Parfit (1987) famously argued that to be rationally concerned about a future individual, one does not need to be numerically identical with that subject; all that matters is if enough direct connections persist between a subject now and a future one. Although identity and psychological continuity generally fall together, there might be (fictional) fission cases, where I today should rationally have an equally strong concern for two future subjects, because both would be psychologically continuous with myself to exactly the same degree.

McMahan, in contrast to that, specifically wants his account to be applicable to all kinds of animals as well. However, a person and a frog (presuming for the moment that the frog is, however weakly, psychologically continuous with himself) do not simply differ with regard to the *quantity* of persisting psychological connections. Direct psychological connections (persistent beliefs, intentions, memories etc.) vastly differ from one another. Having a persistent belief through a (short) period of time, for example, requires much less with regard to the psychology of the individual in question than the implementation of a plan, for instance, that involves the strategic ordering of several action sequences over an amount of time. Plausibly, the mental world of a person contains kinds of direct psychological connections that the frog does not even have access to, or only to a much lesser degree. In aiming to give an answer to the question whether and to what extent animals form psychological unities, these qualitative differences need to be credited.

Both Parfit and McMahan acknowledge that some direct psychological connections are more important with regard to psychological unity than others. Parfit states that "[m]ore weight should be given to those connections which are distinctive, or different in different people" (Parfit 1987, 515), which I take to refer to those connections that play a particularly important role in the context of a person's unique biography or narrative, such as episodic memories or character traits.

McMahan goes further than Parfit in that he is interested in the *formal* type of connections that result in a deeper psychological unity. Consequently, he gives a much more nuanced picture of the complexity of psychological connections and their significance for psychological unity than Parfit. He differentiates between two types of direct psychological connections:

> Some psychological connections are constituted simply by an earlier and a later manifestation of the same mental state – for example, a belief or disposition of character[33] that exists at time t_1 and continues to exist later at time t_2. Other psychological connections are constituted by mental states that occur at different

33 By referring to the disposition of a character persisting through time, I believe that McMahan has in mind something similar to what Simmons (2010, 2006) terms *dispositional desires*, i.e., desires that the individual continues to possess over time. Insofar that a particular animal typically enjoys to play, i.e., we would expect her to form the desire to play under appropriate circumstances, she can be said to have a dispositional desire to play, even if she does not have this desire *right now* (because, say, she is tired). Note, however, that dispositional desires can either be retention-typed or internal reference-typed connections in the sense defined below. Thus, the time-relative interest account can accommodate Simmons' claim that desires can be non-fleeting and persistent and are relevant regarding an individual's interest in continuing to live, while also insisting

> times but contain some internal reference to one another – for example, an experience and a later memory of that experience, or the formation of a desire or intention and the later experience of the fulfillment of that desire or intention. (McMahan 2002, 74)

I will call the first type of these direct psychological connections *retention-typed connections* and the second one *internal reference-typed connections* or, short, *retention* and *internal reference*.[34] Unfortunately, McMahan does not elaborate on the nature of both. First and foremost, it remains unclear from the definition above and other passages in the text whether internal reference-typed connections necessarily require the individual to be aware of this internal reference. I will argue below that they do, because only then it is comprehensible why internal reference-typed connections should be so essential for psychological unity. Yet, internal references between two mental states can occur in various forms.

Take, first, the example of an internal reference-typed desire. McMahan's insistence on the "*experience* of the fulfillment of [a] desire or intention" (McMahan 2002, 74; my emphasis) suggests that the internal reference does not merely hold between the formation of a desire and the later act – even though another passage in the text suggests otherwise[35] –, but between the formation of the desire and the later *awareness* that a desire, that has been formed earlier, has now been satisfied. As human persons, we typically form a desire at some time t_1 and later either remember the formation of that desire at time t_2, when the desire is being satisfied (or thwarted), or, if we do not remember the exact circumstances around the formation of the desire, we at least consciously know that we have had this desire for some time which is now, at t_2, being satisfied (or thwarted).

However, this does not mean that the individual must be conscious of the entire period she has been having this desire. An individual may sustain a desire through time and be aware of its fulfillment at t_2, but still think that she

---

that there is a basic difference between retention-typed and internal reference-typed psychological connections.

34 Retention-typed connectedness correlates with what Marc Slors calls *P-connectedness* (P for preservation), internal reference-typed connections constitute, in his words, *N-connectedness* or *N-continuity* (N for narrativity). N-continuity is both a form of connectedness and continuity, as it is transitive (Slors 1998, 71).

35 The relevant passage suggests that an internal reference exists in the case of "acts that fulfill desires or intentions formed significantly earlier" (McMahan 2002, 74). Taking McMahan's above definition seriously, however, which states that internal references must occur between two *mental* states, it should be clear that McMahan cannot refer to the act *per se* that fulfills the desire.

just formed the desire without knowing that she has actually held this desire for a significant period. Think of a hypothetical animal called Pooch[36], who gets very thirsty during a long walk. Assume that Pooch forms the desire for water and sustains this desire during the next two hours. The moment he enters the house, he directly heads for the bowl and satisfies his thirst. We can acknowledge that Pooch has retained a desire through time and also that he was aware of having had this desire when he satisfied it. But we do *not* need to further assume that he was *aware of the length of time* of having this desire. Pooch does not necessarily think something like "Finally! I have been so thirsty for so long". Thus, an internal reference could, in principle, stretch over an extremely short period, even though the desire itself may have actually been present for a much longer time already.

Yet, the same story of thirsty Pooch can be told somehow differently by help of a retention-based desire. Imagine that Pooch forms a desire for water at time t_1 and sustains this desire through time. At time t_2, that is, the moment the desire is satisfied, it simply *disappears* from consciousness, without Pooch remembering the earlier formation of the desire or being conscious of having had this desire for a while. This would be an example of a retention-typed desire; it is sustained through time, but there is no experience of its satisfaction.

Although McMahan's examples for internal references in the passage above are all backward-looking in that the subject either remembers a past experience or the formation of an intention, internal references can be forward-looking as well. Analogous to a desire, an intention can be sustained through time and then vanish the moment the goal has been achieved, or only refer to the very immediate future. More complex internal reference-typed forward-looking connections, however, allow an individual to anticipate a future mental state, for instance, the satisfaction of a presently held desire or the formation of a future desire.

3.3.2.3 Strong Psychological Unity

It was argued that two types of direct psychological connections should be kept separated. Internal reference-typed connections are "more deeply unifying psychologically" (McMahan 2002, 82) than retention-typed connections. Until now, it remains to be explained why McMahan ascribes internal references such an important role. I will propose below that the reason why internal

36 Although Pooch bears resemblance to a non-hypothetical dog, of course, I want to stress that I do not intend to make any empirical claims about the cognitive capabilities of real canines here. The example of a dog-like animal was taken here solely for reasons of graphic quality.

references are more important with regard to an individual's degree of psychological unity, is that McMahan has in mind those internal reference-typed connections that reveal the capacity to mentally represent scenarios other than the currently perceived or "to transcend the temporally and spatially local" (ibid.), in his own words. This ability goes beyond the limited temporal scope of Pooch from the example above and is necessary for a being to be able *at all* to form a strong psychological unity.

McMahan's emphasis of internal references becomes comprehensible by recapping what strong psychological unity for him amounts to. Strong psychological unity involves, for instance, long-range (internal reference-typed) desires, plans for the more distant future, sacrifices, personal growth or the ability to learn from one's past. The mere *retention* of mental states through time, thus, cannot result in strong psychological unity. Internal reference-typed connections, with the exclusion of those mental states that refer to the immediate past or future, imply the ability to transcend the here and now in McMahan's sense, because the individual is required to consciously anticipate or represent a future (mental) state, that is, a state that differs from what she currently perceives. This will hopefully become clearer in the remainder of this chapter.

McMahan is sure that the ability to mentally transcend the temporally and spatially local is what "raises us above the level of animals" (ibid.), thus representing the main anthropological difference[37]. As a result, "animals generally lack long-range desires; hence they do not consciously plan or make sacrifices for the sake of the future" ibid. 198). Furthermore, the potential for advanced narrative unity, which strongly contributes to the degree of prudential unity relations, as the example of the girl who had initially planned to become an internist, on the one hand, and the volatile person, on the other hand, showed, "is entirely absent" (ibid. 197) in the mental lives of animals. Instead, he pictures animals as follows:

> There are no projects that require completion, mistakes that demand rectification, or personal relations that promise to ripen or mature. Rather, as Aldous Huxley once put it, "the dumb creation lives a life made up of discreet and mutually irrelevant episodes." Each day is merely more of the same. As an animal continues to live, goods may continue to accumulate in sequence, but the effect is merely additive. There is no scope for tragedy – for hopes passing unrealized, projects unwillingly aborted, mistakes or misunderstandings left uncorrected, or apologies left unmake. (ibid. 197)

37 The expression *anthropologische Differenz* derives from Markus Wild (2010).

Incapable of all those internal reference-typed connections that contribute to a strong psychological unity, animals are only weakly psychologically unified. Even more severe, they do not even have the *potential* to develop strong psychological unities. Animals, on this view, are stuck in time. Even though McMahan does not explicitly say so, the transition from weak to strong psychological unity, therefore, seems to be marked through a threshold on this view, defined through the ability to mentally transcend the here and now.

However, McMahan also states that in animals "there is *little* internal reference among the earlier and later mental states: *few* memories of particular experiences, *few* acts that fulfill desires or intentions formed significantly earlier, and so on" (ibid. 74; my emphasis). This speaks against an *a priori* exclusion of animals from the realm of internal reference-typed connections. How does this match my above suggestion, then, that McMahan seems to exclude animals right away from the potential to form strong psychological unities?

By help of Pooch, I sketched an internal reference-typed connection that does not require the ability to mentally transcend the temporally and spatially local. I interpret McMahan so that he allows animals to possess internal references in this limited form. Pooch may very well be able to pre-experience the feeling of water on his tongue as he heads for the bowl. And he may also be aware that he satisfies a need for water while drinking. But these experiences, under this reading, refer to the very immediate past and future, respectively, and they are limited to the dog's current perception of the world around him. This would mean that Pooch is unable to imagine a scenario other than which he currently perceives. To pre-experience the water on his tongue, for instance, Pooch still needed to *see* the water bowl. In this sense, he would still be stuck in the here and now.[38]

This raises the question whether McMahan's position is indeed correct, that is, an empirically valid description of animals' mental lives. If it was shown that (some) animals, contrary to McMahan's hypothesis, do make sacrifices in anticipation of future pay-offs, actively pursue long-range desires and plan for a future situation, this implied that they were much stronger psychologically connected to their future selves than McMahan suggests. Rather than excluding them from the realm of those significant internal reference-typed

38 McMahan even assumes that most goods in the lives of animals "tend to arrive unbidden and indeed unanticipated, for the absence of prior desire" (McMahan 2002, 197). This suggests not only that animals, in his approach, are scarcely aware of their desires at the time a future good manifests itself, which would mean that they have only few internal reference-typed desires, but also that they barely form desires *at all*. If desire is understood as action-motivating, this would mean that animals are not so much actively pursuing their goals, but rather passively receiving pleasures and displeasures.

connections that require the ability to mentally transcend the here and now, some animals had the potential to form strong psychological unities after all. In accordance with the time-relative interest account, animals would be harmed more strongly through their deaths than McMahan grants them to be.[39]

3.4 On Being Stuck in Time

The last chapter delved deep into the nature of the prudential unity relations, determining the weight that the amount of future good the individual loses through her death should have. The stronger the prudential unity relations bind an individual to her future self, the weightier is, *ceteris paribus*, the loss she suffers from death. The time-relative interest account (TRIA) always evaluates a token death, that is, a specific incidence of an individual's death at a given point in time. In determining whether an individual can be harmed by death at all, however, is to ask for what necessarily and sufficiently constitutes rational egoistic concern about the future, providing, in turn, the basis for the prudential unity relations.

3.4.1 *The Psychologically Unconnected Animal*

Intuitionally, one might think that to be concerned about a future event, the future individual this event will happen to must, at the very least, be identical with one's present, concerning self.[40] McMahan shares Derek Parfit's view, however, that in order to be rationally egoistically concerned about one's future, it is enough to be psychologically related to that future subject *in the right kind of way*. Personal identity is not necessary (although it comes along, normally). With regard to what constitutes the nature of this psychological relation, McMahan dissents from Parfit. While Parfit holds that psychological connectedness and (strong) psychological continuity are relevant, McMahan, somewhat surprisingly, argues that the mere continuous capacity

39 In combination with the objections on and suggested modifications of the first element of the TRIA, that have been discussed in chapter 3.2.2, this harm might become even greater.

40 Accordingly, Aaron Simmons concludes that an approach that focuses on the future goods or opportunities for fulfillment, that death deprives an animal of, fails. He argues: "Unlike the case of persons, it is unclear whether animals have the capacity to care about their future opportunities. Having this capacity would seem to require having the capacity to conceive of oneself as a persisting individual and [...] it's dubious whether animals have this capacity. If animals lack the capacity to care about their future opportunities, then it's unclear why their future opportunities would be good for them, or why the foreclosure of these opportunities would harm them" (Simmons 2010, 385).

of consciousness is necessary and sufficient for rational egoistic concern. Why does he so rigorously dissent from Parfit here?

> I suggest that the basis for an individual's egoistic concern about the future – that which is both necessary and sufficient for rational egoistic concern – is the physical and functional continuity of enough of those areas of the individual's brain in which consciousness is realized to preserve the capacity to support consciousness or mental activity. Usually the functional continuity of these areas of the brain involves broad psychological continuity, but in the very earliest phases of an individual's life and in some instances near the end, the same mind or consciousness persists in the absence of any degree of psychological connectedness from day to day. (McMahan 2002, 67f)

To fully understand this passage, several of its concepts must be explained. McMahan separates physical, functional and organizational continuity of the brain, where "the brain" here refers to those areas of the brain where consciousness is realized. *Physical continuity* "requires either the continued existence of the same constituent matter or the gradual, incremental replacement of the constituent matter over time" (McMahan 2002, 68). This understanding of physical continuity allows for a gradual replacement comparable to the famous example of a ship, that is composed of many pieces of wood, which are gradually replaced over many years, or the process of cellular turnover as it occurs, for instance, in the skin. *Functional continuity* describes "the retention of the brain's basic psychological *capacities*" (ibid.). *Organizational* or *structural continuity*, finally, "involves the preservation of those configurations of tissue that underlie the connections and continuities among the *contents* of an individual's mental life over time" (ibid.). Psychological connectedness and continuity are being enabled by this last form of continuity of the brain.

Parfit holds that psychological connectedness and (strong) psychological continuity are the necessary and sufficient bases for rational egoistic concern (Parfit 1987, 313). An individual, thus, has only reason to care about her future self, if she and her future self are at least weakly connected and if there are overlapping chains of strong connectedness between her present and her future self.[41] Thus, Parfit holds that organizational, but not physical, continuity is

41 As McMahan rightly states, the requirement of strong psychological continuity seems rather odd. To require that direct psychological connections become only relevant if they amount to a specific number (which, moreover, is not argued for), "seems arbitrary to claim" (McMahan 2002, 50); more so, because Parfit holds that concern for one's future occurs in degrees (Parfit 1987, 313).

Of course, Parfit's psychological account could easily be revised: rather than holding that psychological connectedness and strong psychological continuity were sufficient bases for egoistic concern, Parfit could accept that *any kind of* psychological continuity, necessary for egoistic concern to be rational. In non-hypothetical cases, functional and organizational continuity rest upon, or are maintained by, physical continuity. As this thesis is concerned with non-hypothetical cases only, we can leave the question aside, whether organizational or functional continuity could theoretically persist without physical continuity.

McMahan now dissents from Parfit in that he holds that organizational continuity, which enables psychological connectedness and continuity, is not required for the ascription of a rational egoistic concern. His main reason to do so seems to lie in his intuition that it is still appropriate for an individual to be concerned about her future in cases, in which there are no psychological connections to that future individual whatsoever. One example is the case of an Alzheimer patient in the early stages of the disease, who is concerned about himself transforming into an *isolated subject*, a "merely conscious entity", "to whom he will not be related even by the weakest strands of psychological connectedness" (McMahan 2002, 55), but who can feel pain. What matters for the ascription of a minimal degree of rational egoistic concern is the continuous *capacity* for consciousness, but not the contents of his mental life over time and, therefore, neither psychological connectedness nor psychological continuity. Nonetheless, these two relations are extremely important with regard to the *degree* of this egoistic concern:

> [S]ufficient physical and functional continuity of the areas of the brain in which consciousness is realized in order for those areas to retain the capacity to support consciousness [...] [are] both a necessary and a sufficient condition of a minimal degree of rational egoistic concern. Beyond that, the degree of egoistic concern that it is rational to have about the future may vary with the degree of physical, functional, or organizational continuity in the brain (or, to be more exact, those areas of the brain in which consciousness is realized). All three forms of continuity matter; and variations in the strength of any of the three are manifest, in practice, in variations in the degree of psychological unity within the life. (ibid. 79)

---

i.e., broad psychological continuity, should suffice. This new version of the psychological account would do justice to our intuition that an Alzheimer patient in the early stages of the disease is justified in being egoistically concerned about his future self at the final stage of the disease, because he and his future self are broadly psychologically continuous. In a similar way, the modified psychological account could ascribe minimal rational egoistic concern to an infant. Despite these merits of the modified version, McMahan's criticism is directed at the idea that psychological connectedness and continuity, regardless of the latter being strong or broad, determine minimal rational egoistic concern, to begin with (McMahan 2002, 50f).

The continuous capacity for consciousness makes all the difference, according to McMahan, between an individual with the capacity to have an interest in his continued existence and a being without.[42] With regard to animals, this implies that even a wholly psychologically unconnected but sentient being has a rational egoistic concern in continuing to live, albeit an absolutely minimal one. McMahan illustrates this by help of an example of a psychologically unconnected but sentient animal, which will be called Psuco here and hereafter. Psuco's mental life entirely consists of a stream of consciousness. Psuco has mostly extremely pleasant experiences, but because he does not even possess the ability to form desires or intentions, he cannot act in any goal-oriented way. Thus, his experiences must be passively received instead of being actively pursued. Psuco, thus, has no intentions, desires or anything else that might build up a psychological relation between himself now and himself in the future (ibid. 75).

Despite his limited capacities, McMahan assumes that Psuco's capacity to enjoy goods provides a reason to care about the being's life *for his own sake* (ibid.). Yet, as Psuco is not psychologically connected to himself, McMahan also seems to hold that the reason for ascribing egoistic concern to him must stem from an impersonal utilitarian perspective. Insofar as the pleasant experiences Psuco will make in the future have value, we have a reason not to kill him, but not because Psuco himself is valuable (ibid. 76). Consequently, McMahan accepts that psychologically unconnected but sentient animals are, what Peter Singer famously called, "replaceable". If it can be made sure that there will be another creature who will make similar experiences, it does not matter if the first creature lives or dies.

This so-called *replacement* or *replaceability argument* has been repeatedly and in length argued to be incoherent elsewhere (e.g., Singer 2015, Singer 2011, Višak 2015). A thorough investigation of this complex topic goes beyond the scope of my research question, however, so I will only point towards one major problem of the replaceability argument, which becomes particularly evident against the background of what Marc Slors (1998) calls a *basic narrative*. I will argue that McMahan's will to nonetheless ascribe psychologically unconnected individuals with an interest in their continued lives becomes much more plausible, in fact, if they are not being regarded as replaceable.

42 Needless to say, of course, that an individual with the *capacity* to have an interest in continuing to live does not always actually possess this interest, if her future looked miserable through and through. Note, however, that even such a miserable being would still possess an egoistic concern about her future, because her egoistic concern does not vary with the prospect of good but only depends on a continuous capacity of consciousness.

Even though McMahan explicitly states that Psuco "lives entirely in the present" (McMahan 2002, 75), he constantly has consecutive perceptions and also sees the movement of a second hand on a clock as continuous motion. Thus, he seems at least to be able to connect these consecutive perceptual contents and make some sense of the physical world around him and his location in it; in short, he must possess what Marc Slors calls a "diachronic unity of consciousness of perceptual contents" (Slors 1998, 72) or *perceptual continuity*, for short. In this sense, an individual's successive perceptions provide him with a basic narrative, where narrative here is not to be understood in a constructivist or teleological sense, but rather in a holistic sense, denying that there can be isolated individual perceptions, thoughts and so on. Under the assumption that experiences cannot be divided from the subject itself, such a perceptually continuous being cannot be replaceable, however. In Slors' words:

> [A] painting must be seen from a particular point of view, revealing the position of the body in the room. Our perceptual contents are not just pictures (sounds, etc.) of the world. They are lived experiences of a situated body that expects her visual inputs to change with the turning of her head, that expects sounds to become louder as she approaches the source, etc., and that questions the veridicality of her perceptions or accepts additional assumptions as soon as these expectations are frustrated. (Slors 1998, 74)

A defender of the replaceability argument could respond that not the experience itself is considered to be replaceable through the experience of a second individual, but the extent of well-being or happiness the experience causes. This, however, is a very different approach, which, importantly, should be applicable to all kinds of beings (including persons). I assume this to be a minor position[43], so I will not get into further detail. Rather, it will be supposed that if a psychologically unconnected but sentient animal is perceptually continuous, it is not replaceable. Thus, I will commit myself to a holistic rather than an atomistic view of mental lives, which argues that experiences are related to the subject itself and, thus, cannot be considered as detached entities.

In fulfilling the basis for rational egoistic concern, Psuco typically also has a *time-relative interest* in continuing to live. This is so, because McMahan has a very broad definition of time-relative interests, where "[o]ne's present time-relative interests are what one has egoistic reason to care about *now* (or, in the case of a non-self-conscious being incapable of being egoistically concerned

43 This seems to be the position Peter Singer leans towards now. Thus, he recently argued for the replaceability not only of merely sentient but non-self-conscious beings, but all kinds of conscious beings, including persons (Singer 2015).

about the future, its present time-relative interests are what a third party would have reason to care about for the being's own sake now)" (McMahan 2002, 80). Even though Psuco is unable to be egoistically concerned about his future himself, he can be ascribed with a rational egoistic(-like) concern about the future and has, if his (net) amount of future good is positive, a time-relative interest in continuing to live.

Psuco, *per definitionem*, is not psychologically related to his future *at all*. All he has is a stream of consciousness. For McMahan, this lack of psychological relatedness is exactly what makes the animal replaceable. Yet, if a being is replaceable and everything worthy about it *entirely* lies in the kind of experiences it makes (but not the being itself), it seems strange to say that the being himself should have an interest in continuing to live or that the being should be ascribed with such an interest *for his own sake*. The replaceability argument rather leans to the claim that *someone*, from the point of view of the universe, has a hedonistic interest in bringing more pleasure into the world, but surely not the individual himself. Abolishing the replaceability argument in favor of a holistic view of experiences that accepts that a psychologically unconnected animal is minimally connected to his future self via perceptual continuity, may actually better fit the time-relative interest account, therefore.

So far, Psuco has been introduced as hypothetical, but the interesting question is, of course, whether psychologically unconnected but sentient animals[44] *de facto* exist. It should be emphasized that these can only be extremely cognitively simple ones, if McMahan's requirement of absolute psychological unconnectedness is taken seriously. Remember that the pool of direct psychological connections does not only include such complex relations as the one between the earlier formation of a desire and the later conscious experience of its (non-)satisfaction, or the experience of an event and the act of (episodically) remembering it, but also contains more trivial forms, such as the retention of a belief[45], desire or a character trait through time.

44 Interestingly, there are non-hypothetical human cases coming at least close to an existence in the moment. Clive Wearing, e.g., a British piano-player and musicologist, lives a life almost completely locked in the present due to an extreme form of retrograde and anterograde amnesia. Unable to episodically remember anything that happened more than 30 seconds ago, he repeatedly finds himself in the situation of having just emerged into consciousness (see, e.g., McMahan 2002, 76f). Another, yet not that severe, case of episodic amnesia was Henry Gustave Molaison, having become known as patient H.M., representing to the day the most famous and most well-researched amnesic patient ever (for an overview, see Augustinack et al. 2014 and Corkin 2014).

45 One might argue that Psuco, being able to make sense of the motion of the second hand on the clock, already retains a belief through time. Analogously, perceptual continuity seems to necessitate the capability to hold beliefs about the external world. Seemingly,

On the first glance, fish or invertebrates might look like good candidates for real-life sentient but psychologically unconnected animals. Yet, findings on the impressive cognitive capabilities of fish and insects, first and foremost, honey bees, has been casting serious doubts on this view (see, e.g., Wild 2012, Segner 2012, Tetzlaff and Rey 2009), even though, of course, it is highly doubtful whether insects fulfill the requirement of sentience. This is not to argue that there may not actually be certain animals falling under the category of psychologically unconnected animals in McMahan's understanding, but they are possibly scarce.

What can be said with considerable certainty is that those animals in the focus of the ethics of killing (e.g., rearing animals, such as cows, pigs, chicken and fish, as well as most zoo animals and pets) are psychologically connected to their future selves to some degree, with the possible exclusion of crustaceans. As this thesis is particularly concerned with the more complex forms of psychological connectedness and continuity and the transition from weak to strong psychological unity, the question whether there in fact exist psychologically unconnected but sentient animals in the sense defined by McMahan will be left unanswered. In the following paragraphs, it will be assumed that the vast majority of morally relevant animals are not like the hypothetical creature Psuco. They are *somehow* bound to their future, but the quality and extent of this psychological relation might still be very different from ours. Whenever referring to *animals* here and hereafter, therefore, it will be assumed that they are at least minimally psychologically related to their future.

That (most) sentient animals are psychologically continuous with their future selves is also acknowledged by McMahan. However, he also denies that animals can mentally transcend the here and now. If this rather vague term is interpreted to refer to the ability to engage in direct psychological connections that take the form of internal references to the non-immediate future, then it becomes evident why McMahan denies animals the abilities to have long-term desires, make plans, prepare and make sacrifices for the future, experience regret and so on. Animals, McMahan argues, are stuck in time, with each day constituting a detached episode, unrelated to the days that have already passed or are yet to come (McMahan 2002, 194ff). Furthermore, their alleged deficit to engage in internal reference-typed connections of the right kind, that is, to the more distant future, excludes animals right away from the mere *potential* to form strong psychological unities.

therefore, McMahan has in mind the sustainment of beliefs over longer periods of time than just the few seconds he allows Psuco to have.

3.4.2 *Who is Stuck in Time – the Animal or the Philosopher?*

> Consider the cattle, grazing as they pass you by: they do not know what is meant by yesterday or today, they leap about, eat, rest, digest, leap about again, and so from morn till night and from day to day, fettered to the moment and its pleasure or displeasure, and thus neither melancholy nor bored. This is a hard sight for man to see; for, though he thinks himself better than the animals because he is human, he cannot help envying them their happiness. (Nietzsche 1876/1983, 60f; cited in Hoerl 2008, 486)

McMahan is far from being alone in arguing that animals are stuck in time.[46] The view of the atemporal animal has a long tradition, whose traces can still be found in numerous recent philosophical, psychological, prosaic and other writings. Animals, on this view, do not know that they have a past and a future and therefore are unable to think about events or scenarios they experienced in the past or which might happen to them in the future, which entirely confines them to a life in the here and now. Whether this should give human persons a reason to pity or rather envy them has been up to debate, however.

The German philosopher Friedrich Nietzsche, for instance, was certain that an animal life was, at least in this respect, superior to a human existence, as the above quotation exemplifies. Others, such as psychiatrist and novelist Irvin Yalom as well as the social psychologists Sheldon Solomon, Jeff Greenberg and Tom Pyszczynski, believe that even though the awareness of ourselves as existing in past, present and future also causes certain anxieties and a profound fear of death, it also leads to certain positive and enlightening experiences (Yalom 2011, Solomon et al. 2015). As the authors also argue that this specific self-awareness is an exclusively human capability, the negative as well as positive experiences coming along with it are argued to be inaccessible to animals. In spite of disagreement regarding the (non-)preferability of a mental existence in the present, thus, most writers have taken it as evidently true that animals actually *are* entirely living in the present moment.

Especially during the last two decades, however, this long-held view has come under serious attack, with numerous comparative cognitivists attempting to provide empirical evidence for animals being able to remember as well as imagine events from their past and their future. In the face of the sheer bulk

46 For a detailed description of how McMahan depicts the mental life of animals, see McMahan 2002, 174–185.

of empirical studies, stating with utter conviction that at least some animals are in possession of planning and other complex cognitive abilities, one might get the impression that the view of the stuck-in-time animal has become obsolete, representing an antiquated, philosophical position like the one of 'man the tool-maker', which came to a shuttering end, when Jane Goodall observed tool-manufacturing and -use in chimpanzees. Could it be, after all, that it is not the animal, who is stuck in time, but the philosopher?

McMahan, like many other writers before and after him, simply presupposes that animals actually are stuck in time or, in his words, cannot transcend the temporally and spatially local. In writing the *Ethics of Killing*, thus, he did not consider the rich empirical data on animal cognition. On the same time, others, who are familiar with the empirical literature or even have conducted behavioral studies with animals themselves, continue to argue in line with the traditional view that the mental lives of animals are stuck to the present moment. Their opponents tend to respond to these objections with more studies, sometimes paired with the claim that their results have simply been misread by their counterparts or that to hold that animals are stuck in time is absurd. In sum, even in the face of many empirical studies on *mental time travel* and other relevant abilities of animals, there still is considerable dispute on whether or not animals should be considered to be stuck in time. It will be argued below that this situation is mainly due to three related reasons.

Often, it is not entirely clear, which abilities exactly an individual is accredited with or denied of, respectively, by the two sides of the debate. The expressions *stuck in time* or *mental time travel* have been used much too ambiguously, with different authors emphasizing different things of what a being with the ability to transcend the present can or cannot do. Further, and related to this point, the classification between an existence stuck in time and its counterpart is too rough to be of real use. A being is not *either* stuck in time *or* a full-blown mental time traveller. Applied to McMahan, this alludes to the similar claim that an individual is not *either* unable to transcend the temporally and spatially local *or* an autonomous person with life-goals, values and principles, possessing long-term desires and, to exaggerate, a detailed project plan for the next ten years. In fact, I will argue in the remainder of this thesis that the space between these simple dichotomies can, and should, be filled with several more fine-grained conceptual classifications.

However, as a third and final point, the mere specification of concepts may not be enough in this context. At some point during the debate, the question inevitably comes up what a concept's 'right' definition should be, that is, a definition that sufficiently captures the essence of the concept, but on the same time does not pose unnecessarily high cognitive demands on behalf of the individual, so that some candidates will be wrongfully excluded right away. This methodological aspect will be discussed in more detail in chapter 4.4.1.

That the scientific community around future-oriented cognition in animals is in desperate need of a conceptual clarification, has been repeatedly stated in the last years, more recently by Cochet and Byrne (2015). *Mental time travel* has become a "catch-all phrase" (Klein and Steindam 2016, 146). Stanley Klein has repeatedly urged to refine the concept of mental time travel in general and of future mental time travel (FMTT) in particular and to be precise in its application. Together with Chloe Steindam, he states:

> [W]e need to be more analytically precise and nuanced about what it is we refer to when we use the label *FMTT*. [...] Is it our ability to construct personal personal future scenarios, to construct public future scenarios, to plan for future contingencies, to anticipate events, to maintain our sense of personal continuity, to make future-oriented judgments, muse about future contingencies, to predict our future feelings, attitudes, and beliefs? (Klein and Steindam 2016, 146)

In an attempt to clear the field, Michaelian et al. (2016) recently proposed to regard *future mental time travel* as umbrella term, integrating any mode how thinking about the future can be conducted. In chapter 4.3, I will propose, in contrast to Michaelian and colleagues, to re-tie mental time travel closer to its initial meaning and rather use future thinking, foresight, prospective cognition or future-oriented cognition as umbrella terms.

CHAPTER 4

Planning: Clarification of an Ambiguous Concept

> If someone asked me whether I personally believe that the other animals are more like human beings than most people suppose, or whether I believe there is some form of deep discontinuity between humans and the other animals, I would have to say yes to both alternatives. (Korsgaard 2006, 103)

Because of the inaccessibility of animal thoughts, empirical sciences cannot measure whether an animal simulates, predicts or intends something *per se*, of course, but only if she *behaves* in a certain manner that allows the inference that she is indeed engaging in future thinking. Planning, in contrast, involves a behavioral output in most cases, and is empirically accessible, therefore. Planning also represents an important component of McMahan's understanding of strong psychological unity, as became evident in chapter 3.3.2. An investigation of animal planning abilities is particularly well-suited, therefore, to find an answer to the question whether animals might be capable to form internal reference-typed psychological connections to their non-immediate future after all, and, thus, are able to transcend the temporally and spatially local, contra McMahan's own assessment. For these reasons, the following considerations will focus on animal planning behavior.

Karl Szpunar and colleagues (2014, 2016) separate future thinking into four basic modes, namely, simulation, prediction, intention and planning. *Planning*[47] is defined as "the identification and organization of steps needed to arrive at a specific autobiographical future event or outcome" (Szpunar et al. 2014, 18418). Even though planning, thus, represents an own mode of future thinking, the authors argue that planning typically involves the other three modes as well. Consequently, planning also involves simulation, that is, "the construction of

47 Szpunar et al. (2014) separate episodic from semantic forms of future thinking and, accordingly, differentiate episodic planning, as defined above, from semantic planning. Semantic in this context refers to non-autobiographical, i.e., general or abstract future states in the world. As direct psychological connections in McMahan's definition can only hold between two mental states, this latter form of future thinking does not have to be considered. However, the distinction between semantic and episodic forms of planning and future thinking in general will prove to be significant in chapter 4.3. The dichotomy of semantic versus episodic planning will be defined slightly differently than Szpunar et al. (2014) propose, however.

 DOI:10.30965/9783957437365_005

a detailed mental representation of a specific autobiographical future event" (ibid. 18416), prediction, that is, "the estimation of the likelihood of and/or one's reaction to a specific autobiographical future event" (ibid. 18417) and the formation of intentions, defined as "the mental act of setting a goal in relation to a specific autobiographical future event" (ibid. 18418).

The special appearances and scope of these three abilities should have a strong effect, therefore, on an individual's planning capabilities as well. In particular, simulation and prediction strongly shape the specific way an individual sees the future, that is, how she anticipates and represents (mental) states that will arise in the future. Before turning towards the concept of planning in chapter 4.4 in more detail, I will first suggest a non-exhaustive list of how differently an individual can actually predict (4.2) and simulate (4.3) future events, scenarios and, accordingly, also her own future mental states. In doing so, I will focus on those differences that will prove to be relevant for the degree of an individual's prudential unity relations. As a stronger degree of these psychological relations generally also leads to a stronger interest in continuing to live, according to McMahan, these differences will prove to be potentially morally relevant in a time-relative interest account.

4.1 Dissociating Future Thinking from Future-oriented Behavior

The foregoing chapter presented Jeff McMahan's time-relative interest account of the badness of death and his depiction of the stuck-in-time animal, confined to a mental life in the present. Obviously, but worth noting, the stuck-in-time hypothesis does not entail that animals cannot *behave* in a future-oriented manner. Animals constantly behave in certain ways that will have payoffs in the future. Future-orientation is indeed a core feature of most behavior (Osvath and Martin-Ordas 2014, 1). To engage in a future-oriented behavior, the animal (or the human), however, does not necessarily have to represent a state in the world *as being located in the future*; in fact, it does not even have to represent this future state *at all*.[48] Two examples for such future-oriented behaviors that do not involve any future thinking capabilities on behalf of the animal are *fixed action patterns* and *associatively learned behaviors*, which will be further explained in this chapter.

Fixed action patterns follow a genetically implemented mechanism, triggered by certain (internal or external) stimuli such as seasonal cues or

48 This is similar to what Saidel (2009) calls goal-oriented in contrast to goal-directed behavior.

hormonal changes. They can be quite elaborate, but are relatively inflexible to disturbances. The maternal feeding behavior of the digger wasp (*Philantus traingulum*) has become what could be described as the "Pavlovian dog example" of such a fixed action pattern, due to its vividness and extensive repetition in the cognitive literature. The relevant behavior is the following: as soon as the digger wasp has caught a grass hopper, she brings him to her nest, that is, a hole in the ground (therefore the digger wasp's name), leaves him outside the entrance, enters the hole – seemingly to check if the entrance is unblocked – and afterwards drags the prey inside.

To see whether the wasp's behavior really was a result of the animal's intention to first check free access prior to dragging the prey inside the hole, Henri Fabre conducted a series of experiments, whose findings he first described in 1879, and which can be found in an essay (Fabre 1915/2001; cited in Keijzer 2013). In one of these experiments, Fabre moved the paralyzed grass hopper for a few inches during the short period of time when the wasp was underneath the ground to inspect the nest. When she came out of the hole again, she found the newly located grass hopper immediately; yet, she did not drag him into the hole, but turned around and checked the entrance once more. This game went on for a considerate time; as long as Fabre moved the prey while the wasp was inside the nest, the wasp re-enacted the action pattern over and over again; to be precise, over forty times (Fabre 1915/2001, cited in Keijzer 2013, 10).

This example has become famous for allegedly representing an absolutely fixed and non-cognitive behavior, even though, as Keijzer (2013) reveals, Fabre himself reported ambivalent results. Whereas some individuals indeed were unable to escape the behavioral "loop" (Keijzer 2013, 17), others adapted their behavior after a few repetitions.[49] Thus, fixed action patterns are not to be misinterpreted as mechanistic, as the digger wasp's behavior has often been, nor does this single incident of a fixed action pattern tell us anything about the wasps' behavior or cognition in general. What can be said with some certainty is that the wasp is not (or only after many repetitions) able to intelligently respond to changed circumstances in this example. Its entrance-checking behavior, therefore, might look rather impressive at first glance, and might even seem to prove that the animal plans ahead, but can actually be shown to be quite inflexible.

49 Even though some individuals modified their behavior, they did not use this as newly acquired information to find a short-cut out of the loop on the next occasion, when the experiment was repeated. Instead, they needed roughly the same number of repetitions on all occasions, before they could finally alter their behavior (see Peckham and Peckham 1905, 304f, cited in Keijzer 2013). This suggests that the behavior in question still was fixed and did not represent an instance of learning.

How associative learning should be defined nowadays has kept several authors occupied in the last couple of years and still sparks lots of discussions. There is broad consensus, on the one hand, that one should be careful not to automatically relate associative learning with extremely simple, non-cognitivist behaviors and to rule out associative learning as a possible explanation for complex behaviors right away; an error Cecilia Heyes (2012) calls *association-blindness*. Rather, associative learning occurs in a wide range of species, including humans. Maybe because "research on associative learning kept bad company – it hung out with people who wanted to explain all behaviour using highly impoverished theoretical resources" (Heyes 2012, 2699), associative learning has long been discarded as non-cognitivist, behaviorist explanation, denying that animals have minds. Nowadays, most acknowledge, however, that associative learning is cognitively mediated, even though the learned content can be represented in a rather thin sense. Consequently, even operant conditioning, an instance of associatively learned behavior, is not to be regarded as cognitively unmediated.

On the other hand, a dichotomy of (higher) cognitive versus associative explanations has manifested itself in the empirical literature on future thinking abilities of animals, suggesting that 'cognitive' here is to be understood more restrictively (see also Buckner 2018, 410). In the context of future thinking, an explanation is commonly understood as 'merely' associative, if the individual in question can be shown to having learned an association between two stimuli, where the relation between these stimuli can be, for instance, causal or temporal, but is not conscious to the individual herself. To check, for instance, whether an animal chose a functional tool over a small reward, simply because she learned to associate the tool with a high reward or, in contrast, because she understood that the tool would provide access to a larger reward in a future scenario, is to check for (and rule out) associatively learned behavior.[50]

4.2 How to Predict the Future

I have argued above that to find empirical evidence for animal planning, the possibility must be ruled out that an individual's seemingly planful behavior is due to rather inflexible mechanisms, such as fixed action patterns or associative learning. Even though especially associatively learned behaviors can be

50 Just because a behavior *can* be explained via associative learning, this does not automatically mean that it is also the right (because cognitively more parsimonious) explanation (Heyes 2012). I will come back to the epistemic value of parsimony chapter 4.4.1.

quite complex and should not be misunderstood as non-cognitivist, behaviorist explanations, it is commonly undisputed that associative learning is dissimilar from planning. In understanding what kind of behavior does allow the inference that an animal actually plans ahead, it will be helpful to understand the circumstances that must be fulfilled to infer this from a human agent's behavior. I will discuss this by help of a well-known situation where we as human agents repeatedly engage in the anticipation of a future mental state, namely by going on errands.

Assume that John's fridge is empty and tomorrow is a holiday, so the supermarket will be closed. Anticipating that he will want to eat tomorrow, John goes grocery-shopping.[51] Depending on whether his current desire is qualitatively similar to his future desire or not – where qualitatively similar here means that John is hungry right now and will also be hungry tomorrow – it can be difficult, however, to identify which of the two desires initiated his behavior. If John is very hungry right now, his grocery-shopping can sufficiently be explained with his present desire for food, even though it may be the case that John really *does* act in anticipation of his future desire. From his behavior alone it is hard to know. If, on the other hand, John just had lunch, then his going grocery-shopping seems to indicate that he really acts in anticipation of a future desire.

On the first glance, it seems to make no difference in this story whether John's fridge currently is filled or not. This is not the case, however. In fact, it could be that John does not think about tomorrow at all. Even though he does not feel hungry, one look into his empty fridge might be enough to make John go grocery-shopping, because this is what one typically does when the fridge is empty. This may be an instance of an associatively learned behavior or a heuristic that is being followed. (In fact, it might very well be that most of our grocery shopping can be thus explained.) To rule this kind of routinely, non-foreseeing behavior out, it would have to be shown that John really anticipates a future desire and not just responds automatically to a present situation that *typically* is not desired. How could this be achieved?

---

51 This necessarily very simplified thought experiment cannot account for the fact that agents often have ambivalent feelings in situations like these. It is possible, e.g., to disapprove of the empty fridge and at the same time wish you *would* approve of it, because you are on a diet. Or you could anticipate your desire for food and still not act in accordance with this future desire, because another, stronger, desire tells you to stick to your diet. This refers to the very complex topic of willpower and second-order desires. Whether any animals are capable of forming second-order desires is an extremely interesting question, I believe, which also relates to McMahan's time-relative interest account. For lack of space, I will not be able to get into this issue here, however.

Assume, in a slightly modified version of our example, that John invited friends for dinner and that he plans to prepare a sumptuous meal, for which all the food that is currently stored in the refrigerator will be required. In this case, John should go grocery-shopping, even though he is currently sated *and* in possession of a filled fridge, because his fridge would still be empty by tomorrow. In this case, John's behavior (grocery-shopping) would give us sufficient reason, I believe, to assume that he anticipated both that his fridge will be empty by tomorrow (because he invited friends for dinner) and that he will be hungry tomorrow.

That an individual *acts* in accordance with her future desire is not sufficient, in conclusion, to infer that the individual priorly *anticipated* her future desire, even if her future desire is qualitatively dissimilar from her present one. This is so, because it must additionally be ruled out that the individual's behavior is explainable through a learned association or an otherwise routinely acquired learning process, as, for example, when John goes grocery-shopping, because of a certain state of the fridge (*Whenever the fridge is empty, go grocery-shopping*).

One way to rule this automatic behavior out, is to make sure that not only the future desire but also the external state qualitatively differs from the present. If John just had lunch and his fridge currently is filled with food, then it becomes extremely likely that his behavior (grocery-shopping) is resulting from the insight that tomorrow, when the supermarket will be closed, his fridge will be empty and John will be hungry again. A fixed rule (*Whenever the fridge is filled, go grocery-shopping* or *whenever you are sated, go grocery-shopping*[52]) would not be very sensible. I conclude that cases, in which the individual's current and future desire are qualitatively dissimilar *and* the individual must understand that the future state in the world will also be different from the present state, are particularly well-suited to approach the question whether animals are able to anticipate their future desires.

These considerations lead over to the key point I would like to make in this chapter. Of course, being able to anticipate a future state requires more than the knowledge that the future will be *somehow* different from the present; one must be able to *predict* what is going to happen. This simple but important point has often been neglected in the debate around prospective cognition in animals. Prediction will be defined with Karl Szpunar, Nathan Spreng and

52 It is true that often human beings are not very good in distancing themselves from their current desire, as, e.g., when they tend to buy much more food than they actually need when shopping hungry. It may, therefore, be actually wiser to go grocery-shopping when you are sated. Still, however, this does not mean that you should go grocery-shopping every time (or even just most times) you feel sated. Thus, an automatic following of the rule *Whenever you are sated, go grocery-shopping* would still be inefficient.

Daniel Schacter as "the estimation of the likelihood of and/or one's reaction to a specific autobiographical future event [or] a general or abstract future state of the world" (Szpunar et al. 2014, 18417). Antedating the conclusion that will be made below, however, I would like to add that a mature ability to predict the future is closely related to the comprehension that the future very often is uncertain, because it might unfold in various ways. It is one thing to predict and prepare for the most likely thing to happen next and quite another to predict and prepare for multiple, mutually exclusive future scenarios. This is loosely connected to a causal theory of time which, very roughly speaking, states that two temporal events can be understood via their causal relationship.[53]

As human adults, we can make sense of the idea that that the past and the present shape the future in multiple ways, some of which seem to be more complex than others. Because the way the future is being perceived influences how a future mental state can be anticipated, as well, it is of particular interest for this research project to know how animals think about the future and to what regard their sense of the future might be different from ours. There are numerous ways how a future state could be predicted. Here, the focus will lie on one major difference, namely that between the ability to anticipate future scenarios, on the one hand, and the ability, on the other hand, to additionally represent the temporal-causal relationship between the mentally imagined scenario and current reality. I will focus on these abilities, because they are particularly relevant with regard to the degree of psychological connectedness.

I will proceed as follows. First, an individual might know what is going to happen by having a relatively simple understanding that some event or action produces a certain effect (4.2.1). More complex than that is the knowledge that one event or action may result in one of two or more (equally likely) events. This refers to the idea that there are several possibilities how the future might unfold (4.2.2). These two basic and contrasting ways to see the future will then be argued to be grounded in the dichotomy between merely being able to temporally update one's model of the world and, in contrast, the capability of temporal-causal reasoning (4.2.3). Thereafter, hypothetical and counterfactual think will be further clarified (4.2.4). Last but not least, these abilities will be analyzed with regard to their significance for an individual's degree of psychological connectedness. I will argue that temporal-causal reasoning abilities,

53 See also Christoph Hoerl's and Teresa McCormack's (2016) claim that, in children, the development of an understanding of time is closely linked to their grasp of causal relationships. Animal cognitivists have long been interested in the question whether and to what extent animals have an understanding of causality, but only recently has causality been studied with an eye on its relevance for future-oriented cognition.

enabling hypothetical and counterfactual thinking, increasingly psychologically bind an individual to her future self.

4.2.1 *Causal Understanding*[54]

To predict what is going to happen involves an understanding of the consequences certain events or actions have. This, however, can be more or less cognitively complex. Jim Woodward differentiates between three different causal learners. An *egocentric causal learner* can learn about causal relationships only from her own actions. She is capable to understand, for instance, "that if [she] performs action A, outcome O follows, and that if [she] refrains from A, O does not (assuming no other cause of O is operative)" (Woodward 2011, 17). Put differently, the individual learns that whether or not she performs action A makes a *difference* to the outcome. (This is why Woodward also speaks of difference-making accounts of causation.) A subject with such a very basic causal understanding, who has pushed down an object through a tube, for instance, would expect another object to emerge from the other end of the tube as well, for example.

An *agent causal learner*, in contrast, learns about causal relationships from her own and other agents' actions. From observing another conspecific, for instance, the agent causal learner is capable to understand that she herself (and other subjects) could bring about the same outcome O by performing action A.

An *observation/action causal learner*, finally, is additionally able to abstract the relevant knowledge from observations that do not involve the action of an agent, but occur "naturally" (ibid. 18). The full range of this kind of causal understanding becomes particularly evident by help of an example by Tomasello and Call (1997), that Woodward refers to: suppose that an ape observes an apple falling to the ground as a result of the wind shaking the branches of a tree. If the ape is an observation/action causal learner, she should understand that the movement of the branch caused the fruit to fall and this would enable her to think about alternative ways how to cause similar movements of the tree branches (e.g., by shaking the branches) to obtain more apples (Tomasello and Call 1997, 389; cited in Woodward 2011, 20).

Woodward's three causal learners exemplify that we must be careful in ascribing someone with causal understanding and be clear about what we mean

54 What I have in mind here by causal understanding is not so much a knowledge of the physical properties, mechanisms and forces and their relationships to one another to provide an account of causality as it occurs in the world, but a more psychological account of the different ways how an individual might represent and learn about causal relationships and integrate this knowledge to act in a certain way, similar to Jim Woodward's (2011) interventionist account of causation.

when we talk about it. That an infant expects the toy mouse she drops to come out of the bottom end of the tube, does not yet indicate that she can also abstract this information and apply it to other contexts as well.

Because she understands that if she performs action A, outcome O follows and also that if she does not do A, outcome O will be absent, it might be argued that even an egocentric causal learner can make certain basic counterfactuals and has an understanding of alternative future possibilities. I will argue against this below on the grounds that it should be differentiated whether an individual is merely capable of thinking about alternative scenarios or whether she is able to represent several likely *future* scenarios in understanding that there are several options for the future to unfold.

4.2.2 *Branching Futures*

More challenging than the expectation of what is going to happen based on a knowledge of certain consequences an event or action has, is the knowledge that there are two or more options how the future could unfold. The tube from the example above could be branched, for instance, so that the object could either exit the right or the left tube. This may look like a very simple cognitive demand from our point of view. Interestingly, however, infants younger than 4 years of age as well as great apes seem to considerably struggle with grasping this (Beck et al. 2006, Redshaw and Suddendorf 2016).[55] Thus, the possibility exists that this may point towards a real limitation in animals' and infants' representation of the future.

Using the branched-tube task, Jonathan Redshaw and Thomas Suddendorf (2016) found that all of the participating 2-year-old infants spontaneously prepared for the object's emergence only from one exit, even though they had been made familiar with the functionality of the branched tube by observing the treat coming out of the left or the right exit for several times right before the 12 test trials. Many 2- and 2.5-year-olds eventually covered both bottoms at least once over later trials, but then regressed to the one-exit-covering-strategy, which is consistent with trial-and-error learning.

In contrast, none of the 4-year-olds, once they had used both hands to cover the exits, regressed to the unsuccessful strategy. Most 4-year-olds, in fact, seemed to understand from the beginning that the desired treat, in this case a ball, could either emerge from the left or the right exit by placing both hands under the tube from the first trial on. Performance of the 3- and 3.5-year-olds

55 Whenever empirical studies will be summarized or analyzed, relevant numbers will be presented in numeric form, so that the important information is more readily identifiable for the reader.

was much better than that of the younger participants, but the proportion of 'regressors' was still relatively high.[56]

Additionally, the authors tested 3 chimpanzees and 5 orangutans on the branched-tube task. Like the 2-year-old children, none of the apes prepared for both possible outcomes of the desired object, a grape, on their first trial. Only one chimpanzee named Holly covered both exits over the course of the 12 trials at all, during trials 9 and 11, but regressed to placing only one hand under one of the bottom openings on trials 10 and 12. Her response pattern, thus, is similar to that of the group of the 2-year-old children.[57]

The authors tentatively raise the possibility that the ability to mentally simulate alternative possibilities and their consequences may be uniquely human and only emerges around the age of 3 to 4. These findings are in line with Redshaw's hypothesis that metarepresentation might be a uniquely human capacity, which he had formulated two years earlier (Redshaw 2014). Metarepresentation, according to Redshaw, "allows humans to represent the nature of the relationship between current reality and alternative representations of reality. Without metarepresentational insight, the future cannot be represented *as the future*" (ibid. 1).

On this reading, animals (and young infants) may very well be able to represent an alternative version of reality, which they know is somehow different from the here and now. These uncontextualized representations can even "function as lower-level 'desired world states' that can motivate present action or bias future action in an adaptive manner, without the agent having an explicit notion of the future representational perspective to which these desired world states belong" (ibid. 3). What metarepresentational insight adds to this form of intentional behavior, is an appreciation of the *nature* of the relationship between current reality and the future. Put differently, an individual

56 Redshaw and Suddendorf's results contrast with those by Beck et al. (2006), who found that children consistently solved the branched-tube task even only during their fifth year. According to Redshaw and Suddendorf (2016), these contrasting results may be due to the higher demand of language comprehension in Beck et al.'s study, however (see Redshaw and Suddendorf 2016, 2 for details).

57 Because at least 3 great apes eventually used both hands, Redshaw and Suddendorf (2016) argue that their poor performance cannot be attributed to an inability to synchronize the actions of two hands through mesh. To my knowledge, however, it remains a possibility that the apes were averse to putting both hands through mesh – maybe even more so, if their backs were exposed to open space –, as this significantly limited their mobility. Under this reading, the majority of the tested apes, who consistently failed to use both hands, might simply not have considered the reward to be large enough to put themselves into this potentially uncomfortable situation. In using only one hand, they opted for the second-best, but less risky, strategy.

without metarepresentation cannot recognize that what she represents might be incorrect or that her two successive mental representations in fact refer to two mutually exclusive possibilities of what is going to happen.

With regard to Redshaw's own findings on the branched-tube experiment, this would mean that the young infants and the great apes may very well have a representation of the desired object falling from one of the bottom openings or that they may even have two successive representations of the object emerging from the first and then the second opening – yet, without a grasp of the *relationship* between these representations and the current reality, they do not accordingly prepare by placing both hands under the tube exits.

Redshaw and Suddendorf (2016) tested in a second experiment, whether the great apes would learn to consistently place both hands under the apparatus, when given more trials. They found that the chimpanzee Samantha and the male orangutan Dinar (but not Holly from the first experiment) eventually learned to cover both openings on at least 6 trials in a row. None of them sustained this strategy, however. As the authors argue, "[s]uch a response pattern – an initial appearance of the target behavior after many unsuccessful trials, followed by a brief regression, then reappearance of the behavior and eventually maintenance – is consistent with simple operant conditioning principles rather than insightful behavior" (Redshaw and Suddendorf 2016, 2).

As becomes apparent in the paper's supplemental information, however, orangutan Dinar prepared for two possible outcomes for the first time on trial 16 and kept this strategy until the end of the experiment (24 trials), with only one incorrect exception on trial 17. On the one hand, it might be possible that Dinar simply made a mistake on one trial. If, on the other hand, he truly represented the future outcome as uncertain, then why did it take him 6 observation trials and 15 test trials to gain this insight?[58] For future research, it would be interesting to know whether 5-year-old children, who should be able to understand that one event can have two mutually exclusive possibilities, make similar mistakes on the branched-tube task when tested not on twelve but on 24 or more trials.

Given their general good physical understanding and due to having been able to observe the workings of the apparatus, it is probably unlikely that the infants and the apes simply did not grasp the function of the branched tube. Alternatively, however, it might be argued that apes do very well understand

58 On the next day, when Dinar was photographed performing the task, which was not part of a formal trial, he regressed again to covering only one bottom opening. This strengthens the authors' interpretation that even in Dinar, no real insight into the future's uncertainty had taken place (Redshaw and Suddendorf 2016, supp. mat.).

that there may be two or more possibilities of how the future can unfold, but that they simply did not think about using both hands in this particular task. Thus, one might object that the requirements of the task with its highly artificial apparatus was inapt to make any inferences concerning the animals' grasp of the future.

Given that the animals, however, eventually used both hands, but, importantly, did not consistently retain this strategy, this becomes extremely unlikely. If the animals knew that the object could emerge from either one of the two exits, but simply did not think about using both hands to prepare for both possibilities, then they would have been expected to retain this successful strategy, once they conceived of it.

This does not rule out the possibility that apes and infants will be demonstrated to succeed on other kinds of tasks that require them to prepare for two equally possible outcomes. For instance, they could be tested on understanding that they would be required to store two copies of one tool at two different locations, given the unpredictability of the location at which they would actually need the tool later. If it turned out that apes (and infants) do not succeed on these tasks as well, this would strengthen Redshaw and Suddendorf's (2016) and Redshaw's (2014) hypotheses.

It may be questioned, whether the branched-tube task is not just about a grasp of cause and effect or certain physical properties. Put differently, one could doubt whether a failure to succeed at this kind of task really allows inferences concerning the participants' temporal notions and their understanding of the future, more so, because the period of time between the insertion and the reappearance of the item is extremely short. Especially *because* it only stretches over such a short amount of time, however, it becomes particularly interesting that apes and infants alike considerably struggle on such an easy task. Thus, the branched-tube task does not directly assess the ability to prepare for a non-immediate future, but could rather point towards some individuals' important limitations to do so. Thus, it could very well represent an important first step to investigate these individuals' underlying understanding of time.

In general, this paradigm, which originates from the above mentioned study by Sarah Beck and colleagues (2006), offers an impressively minimalist way to test for children's and animal's understanding of branching futures. More and diverse research will shed light on whether animals may eventually be shown to succeed on these and similar tasks (see also Suddendorf et al. 2017, 2019 for slightly different versions of the task). For now, it would be premature to make any conclusive claims (Redshaw and Suddendorf 2016, 3; see also a review by Seed and Dickerson 2016).

4.2.3 *Temporal Updating vs. Temporal-causal Reasoning*

In predicting the future, very often whole sequence of events, rather than one singular event, are being mentally simulated. Because human adults typically have a large temporal horizon and can simulate multiple possible scenarios of what might happen, their predictions of future states can be quite complex. Depending on how many steps into the future the subject must look ahead to solve a problem, there will very likely be major developmental and comparative differences in performance. As has already been seen in the branched-tube test, infants and animals may perhaps not be able to grasp two sequences of events as mutually exclusive future possibilities, even though they might very well be able to represent these two sequences one after the other.

Christoph Hoerl and Teresa McCormack (2011) now argue that even though very young children (and many animals) can represent events in a temporally ordered way, they seem incapable of thinking about the temporal order of hypothetical events itself. Put differently, this means, in their terminology, that young children and animals do not have fully developed temporal-causal reasoning abilities but merely engage in *temporal updating.*

> In temporal updating, [the individual] has a model of the world (e.g., of where certain objects are located) that changes as and when she receives information about changes in the world. In other words, there will be a sequence of changes to the [individual's] model over time as a result of the [individual] receiving information about each event in turn, without the [individual itself] having to reason about the order in which those events happened. (Hoerl and McCormack 2011, 453)

Hoerl and McCormack's hypothesis is being supported, at least in case of infants, by previous research (McCormack and Hoerl 2005, 2007). One of their studies, for instance, shows that while 3-year-old children could easily recall the order in which two events (one doll placing a hairbrush into a blue cupboard and another doll thereafter putting the same hairbrush into the red cupboard) had happened, they were only able to indicate the hairbrush's present location, when they had been watching the relevant events. When the children's views were blocked so that they could not witness that the hairbrush was first being placed into one cupboard and then into the other, but were only verbally informed about these events, even 4-year-olds could not indicate the hairbrush's location above chance, despite being able to recall the order in which the two events had happened. 5-year-olds, in contrast, were not dependent on witnessing the action sequence for themselves. The authors hypothesize that the 3- and 4-year-olds, unable to update their model of the world in the blocked-view trial, did not yet understand which of the two past events indicated the hairbrush's final location.

Furthermore, McCormack and Hoerl's claims can be neatly related to Redshaw's position that young children do not yet have a grasp of the nature of the relationship between a representation and current reality. Even though the 3-year-olds in McCormack's and Hoerl's study correctly recalled what had happened, they were seemingly unable to understand the temporal-causal relationship between these representations and current reality. When they were allowed to watch the hairbrush being transferred from one location to another, they did not need to bring these two representations into a temporal-causal order afterwards, because they could answer where the hairbrush now was by merely updating their model of the world as time proceeded.

With regard to future event sequences, McCormack and Hoerl propose that young children and animals may be able to simulate a short event sequence in advance of action without being simultaneously able to reason about the order of these future actions. Moving through time, they always only know what is going to happen next, without being able to flexibly think about the order of these events.

> As described here [in Hesslow 2002], a simulation 'runs off' an action sequence, with each simulated action triggering the simulation of the next action due to the perceptual state that is anticipated. What this means, in the first place, is that engaging in this form of simulation does not itself require thinking of the simulated action as occurring in a specific temporal sequence; the [individual] simply ends up simulating them in that sequence. (McCormack and Hoerl 2011, 133)

An individual with fully developed *temporal-causal reasoning abilities*, in contrast, can think about the temporal order in which several events or actions necessarily need to happen to achieve a certain outcome in a task. Children's success with tasks that require temporal-causal reasoning greatly improves between ages 3 and 5 (McCormack and Hoerl 2007, 181). Thus, the authors suggest, "reasoning about the causal significance of the temporal order of events may not be fully developed before 5 years" (ibid. 168).

The representation of branching futures can be seen as a simple instance of temporal-causal reasoning, because it already seems to presuppose the understanding that two events stand in a particular temporal relation to one another, even if this relation is one of simultaneity, not of succession here. Thinking about the manifold ways how the future could unfold, typically involves thinking about multiple mutually exclusive possibilities and also knowing that the order of certain events is dynamic rather than static. Consequently, the capability to think about the causal significance of the order of two or more events is particularly important in those instances, in which the most effective of multiple, mutually exclusive possibilities is to be determined, for instance, when temporally structuring one's errands and appointments or determining the best route through a complex maze.

To sum up, Redshaw as well as McCormack and Hoerl acknowledge that, first, animals and young children may be able to think a few steps ahead in mentally representing a short event sequence (temporal updating), without also being capable of reasoning the temporal-causal relationship between these events (temporal-causal reasoning). Furthermore, these individuals may also be able to represent more than one of these event sequences in succession without simultaneously understanding that these sequences constitute alternative future scenarios (branching futures).

4.2.4 *Hypothetical and Counterfactual Thinking*

Hypothetical and counterfactual thinking, as understood here, presupposes the knowledge that some events, that lie in the past and are unalterable now, had once represented mere future possibilities. Here and hereafter, *hypothetical thinking* will be defined as ability to think about possible future scenarios and events. *Counterfactual thinking*, in this context, refers to the ability to think about what has not but could have happened at some point in the past, because the corresponding events had once constituted future possibilities. A subject with counterfactual thinking abilities is aware after each trial of the branched-tube task that the object could alternatively have emerged from the other exit.

Hypothetical and counterfactual thinking play a major role in the capability to feel regret. John knows that if he had not asked his friends over to dinner, he would not have to go grocery-shopping now, which might make him regret his past decision to invite his friends. As Hoerl and McCormack emphasize, it is even possible to regret a future decision, that is, a decision that has technically not been made yet. What they call *anticipated regret* refers to the thought that a certain action or event will have certain consequences, that will not be desired by one's future self and that will cause a negative emotion, because he will look back to this particular moment and regret the decision he made then, knowing that he could have decided otherwise. Anticipated regret thus is a form of "prospective retrospection" (Hoerl and McCormack 2016, 250). John knows that he can choose between several options what to do next (go grocery-shopping or reading a book, for instance), which will have consequences for how the future will unfold. Consequently, he can also anticipate not one, but several possible future mental states, for instance, that a filled fridge tomorrow will make him feel sated, whereas an empty fridge will make him regret his decision of having read a book instead.

These examples seem to indicate that in order to think about counterfactual and future hypothetical alternatives, one necessarily needs to be able to think of what has been called branching futures. Thus, when Jane can correctly answer the question where the mouse would have emerged, if he had taken the other way (counterfactual) or where the mouse will be, if he takes the other way next time (future hypothetical), it seems that she must be able to understand that at some point in time, there had been or will be two possibilities for the mouse to exit the tube.[59] In fact, things are a bit more complicated, however.

Beck et al. (2006) found that on only 21 and 37 percent of the trials, respectively, the two groups of younger (mean age: 3 years 7 months) and older children (mean age: 5 years 1 month) spontaneously prepared for the two possible outcomes of the toy mouse by placing mats under both bottom openings, so that the mouse would land safely and not 'hurt' himself. Surprisingly, however, the children had no difficulties in correctly answering a future hypothetical question ("What if next time he goes the other way, where will he be?"), making mistakes on only 6 percent of the trials (Beck et al. 2006, 218).

In a second experiment, 3- and 4-year-olds were found to correctly answer the standard counterfactual question ("What if it [in this case, a cotton wool ball] had gone the other way, where would it be?") in 87 percent of the trials and 5- and 6-year-olds in even 98 percent of the trials. Interestingly, both groups performed much poorer when being asked an open counterfactual question ("Could it have gone anywhere else?"), however: 3- and 4-year-olds gave a correct answer in only 66 percent of the trials and 5- and 6-year-olds in 83 percent (Beck et al. 2006, 421).

To sum up, the children were very good at answering standard future hypothetical and counterfactual questions, but performed much poorer when being asked an open counterfactual question or when they were required to prepare for two mutually exclusive outcomes.

In order to account for these results, the authors propose that to answer the open counterfactual question correctly, children needed to mentally

59 There is considerable empirical evidence that counterfactuals pose greater difficulties to children than matched future hypotheticals and that counterfactual thinking evolves later in development – around the age of four – than hypothetical thinking, which seems to be adult-like by three years of age (Beck et al. 2006, 415). It has been proposed that this might be due to the fact that, while both counterfactual and hypothetical future thinking require the individual to imagine a different state of the world, counterfactual thinking requires the individual to additionally disregard her knowledge about current reality.

travel backwards in time to a point when multiple events could have happened, whereas the standard counterfactual question only requires them to imagine an alternative world. They clarify this very nicely by help of an example from a study by German and Nichols (2003). In this study, children were told that Mrs. Rosy is in the garden and opens the kitchen door to call her husband. When she opens the door, the dog escapes from the kitchen into the garden and tramples on a flower, which makes Mrs. Rosy sad.

While the 3-year-old children from the study found it easy to answer "What if the dog hadn't squashed the flower, would Mrs. Rosy be happy or sad?" (short counterfactual chain), they had difficulties to answer "What if Mrs. Rosy hadn't called her husband, would Mrs. Rosy be happy or sad?" (long counterfactual chain). Beck and colleagues suggest:

> [C]hildren could in principle give the right answer to short counterfactual chains by imagining only an alternative world and not relating it to the current world. That is, one can reason simply "dog does not squash flowers, therefore Mrs. Rosy happy." In contrast, thinking "Mrs. Rosy does not call her husband" does not lead to an obvious "Mrs. Rosy happy" or "Mrs. Rosy sad" conclusion. The child *must* think back in time and consider the consequences of this change. (Beck et al. 2006, 423)

A basic understanding of causality, together with the ability to mentally represent an alternative world, is sufficient to correctly answer the short counterfactual chain question here. Similarly, the children in Beck et al.'s study were able to answer the hypothetical and standard counterfactual questions simply by imagining an alternative scenario, whereas the open counterfactual question probably required them to actually think back in time. Analogously, the children could easily answer the hypothetical question even in the absence of wholly understanding that the alternative world that they imagined constituted one of two possible future outcomes. In line with the proposals made by Redshaw (and also Hoerl and McCormack), these results could be interpreted to show that the younger children simply engaged in a mental representation of an alternative version of reality, without also representing the temporal relationship between this representation and current reality.

If, however, the temporal aspect is deemed essential for future hypothetical and counterfactual thinking in that we want to ascribe these capabilities only to those individuals with a grasp of the fact that there are multiple ways for the future to unfold, then the ability to correctly answer standard counterfactual or hypothetical questions is insufficient to provide evidence for counterfactual and hypothetical thinking abilities. The branched-tube task is a valid first measure to investigate basic hypothetical thinking abilities, therefore, because it requires participants not only to mentally represent alternative scenarios, but to additionally represent the temporal-causal relationship between these two representations and current reality.

4.2.5 *The Relevance of Prediction for Psychological Unity*

I will conclude this chapter by explaining to which extent these different ways to predict the future, that have been presented above, are relevant for how much an individual is psychologically connected to her future self in the sense defined by McMahan. It will be assumed that an animal capable of anticipating her future needs or desires[60], is, at least in theory, able to prepare for these. Just like John should go grocery-shopping on the day before the holiday, an animal with planning abilities should cache and store food, even if she does not feel hungry and there currently is plenty of food available. As has been laid out in chapter 3, McMahan confidently claims that animals are unable to plan for a future need. In fact, he seems to regard this ability to represent a crucial threshold, enabling an individual to step out of an entirely present-oriented mental life and, therefore, as a necessary requirement for the capability to form a strong psychological unity to begin with.

An animal with hypothetical thinking abilities, who can anticipate not only one but several probable or possible future worlds and her related alternative mental states, should, in principle, be able to prepare for alternative, mutually exclusive events. Needless to say, McMahan is certain that animals do not possess this kind of forethought as well. More importantly here, however, is that the ability to anticipate not only how the future *will* unfold but the different ways how it *might* unfold, binds a subject even stronger to her future self, because each anticipation of a future possible mental state represents an internal reference-typed connection.

As has been seen, recent empirical studies raise the possibility that animals may indeed not be able to simultaneously prepare for two possible

---

60 Markus Wild (2010, 123ff) separates (at least) two different kinds of motivational states, namely needs (*Bedürfnisse*) and desires (*Wünsche*). A need aims at bodily changes, whereas a desire aims at changes in the world. An example for a need thus is hunger or thirst, whereas an example for as desire could be the motivational state to build a nest for the night. In the empirical literature, needs and desires are not differentiated most of the time; furthermore, it is often difficult, if not impossible, whether a specific behavior aims at the satisfaction of a future need or desire the animal has. Regarding the prudential unity relations, however, this conceptual differentiation is not of utter importance, at least in those cases where the relevant anticipation of a future need or desire suffices McMahan's definition of an internal reference-typed psychological connection into the (non-immediate) future. Needs and desires will not be neatly conceptually separated, especially when directly referring to empirical literature.

future outcomes. In line with the arguments made above, this would imply that animals, just like children younger than five or six, may not be capable of temporal-causal reasoning as well as hypothetical and counterfactual thinking. This possibility is proposed and defended by, for instance, Teresa McCormack and Christoph Hoerl as well as Jonathan Redshaw. The position that animals do not have a grasp of temporal-causal relationships, opens an interesting avenue for future research. For now, more research is needed to strengthen this hypothesis.

Counterfactual thinking should even further contribute to the degree of psychological unity, compared to hypothetical thinking, because it allows one to evaluate one's own path of life and may thus be essential for narrative unity. It also further strengthens the relation of the subject to her future self, because anticipated regret represents a forward-looking internal reference-typed connection, which additionally includes a backward-looking internal reference. McMahan emphasizes that only in the lives of persons there is a "scope for tragedy – for hopes passing unrealized, projects unwillingly aborted, mistakes or misunderstandings left uncorrected, or apologies left unmade" (McMahan 2002, 197). These tragedies seem to presuppose counterfactual thinking, because they all seem to resolve around the common topic of regret. One can regret not having pursued certain wishes and projects[61], but also feel regret in light of unsettled disputes, unresolved mistakes and misunderstandings, and so on. The ability to look back on one's life and regret past decisions seems to be of particular importance for McMahan. As he is convinced that animals do not have the ability to reason about past (or future) decisions or events in this way, he does not believe that animals can engage in hypothetical and counterfactual thinking as defined above.

61 In the quotation above, McMahan emphasizes "unwillingly aborted" (McMahan 2002, 197) projects, that is, projects that the individual was forced to give up. Imagine, e.g., a young man, who feels required to give up his academic studies to care for his younger siblings after his parents' sudden death. Even though it might very well be the case in this example that the man will come to regret not having resumed his studies later, it seems that individuals also occasionally regret things that lie outside their scope of action. One can, thus, approve of one's decision, while also feeling regret (*Bedauern*) in light of other options that were eliminated by one's decision or feel regret in light of an alternative possible scenario, in which one's parents had not died so soon. Regret, thus, may be quite complex. In general, however, it seems closely linked to an understanding that a scenario or a decision, lying in the past now, had once merely represented one of several possibilities of how one's life could have unfolded.

4.3 How to Simulate the Future

As has already been indicated above, there may be two ways to simulate a short sequence of events or actions, either via what Hoerl and McCormack call temporal updating or by temporal-causal reasoning. What remains to be seen, however, is how this difference relates to the commonly made distinction between *semantic* versus *episodic simulation*, looming large in the debate around animal planning.

Two ways how a sequence of future events can be simulated have been differentiated from one another. An event sequence can be simulated either on the basis of more general information and regularities about the world or from a first- or third-person perspective. Thus, a subject could anticipate certain future events happening, predict her future motivational relation based on rather general information she has, define a goal state and act accordingly. Assume that Jane, for instance, anticipates that her guests will eat salad, then lasagna and expect dessert afterwards. She predicts that she will want to serve them dessert, which she has not yet prepared, so she decides to go get some ice cream.

Alternatively, a subject could imagine himself to be in the future scenario, evaluate the situation, define a goal state, mentally travel back to the here and now and act accordingly. John, for example, might imagine himself sitting at the dinner table with his friends, eating first salad, then lasagna, and then imagines his friend Karen asking "So, what's for dessert, John?", at which he feels slightly embarrassed for not having prepared dessert. He could then simulate various scenarios with different kinds of dessert to find out which one will fit best (after having eaten a lasagna, one will probably not be too thrilled to eat a rich creamy chocolate cake), but sooner or later he will mentally travel back to the present and go get ice cream. Of course, many variants of these two described processes are conceivable; for instance, John may only have a short glimpse at the future scene (*no dessert*), instantly mentally travel back to the present and only there define a specific goal state (*get ice cream*). Furthermore, simulations typically involve aspects of both ways.

Most authors would agree that Jane is engaging in semantic future thinking, whereas John is exercising episodic future thinking. In the process of the debate around prospective cognition in animals it has become less and less clear, however, what *exactly* the difference between both amounts to. This is at least partly due to the multifaceted nature of episodic future thinking. Depending on which of the various facets are regarded as *essentially* episodic – that is, constituting the essence of what makes a thought episodic in nature –, different notions of episodic future thinking have emerged. Ideally, these different ideas complement one another, but they might also contradict. Moreover, a growing literature on the phenomenology of episodic future thinking strengthens the common intuition that even only among human persons, there may actually be great individual differences with regard to episodic thinking, culminating in the thought that *my* way to simulate the future may be quite different from *yours*.[62]

The struggle to define episodic in contrast to semantic cognition has been going on for over thirty years now and I will not attempt to solve it in this thesis. Rather, I will suggest that there have evolved two main strands of how episodic future thinking is to be defined in contrast to semantic future thinking and that, consequently, they should always be clearly indicated as to avoid misunderstandings. In doing so, it will be helpful to give a historical outline of the debate.

4.3.1 *The Struggle to Define Episodic Future Thinking*

The distinction between semantic and episodic future thinking goes back to Endel Tulving's classification of the declarative memory into a semantic and an episodic memory system. Despite its intuitive plausibility with regard to the past – knowing vs. remembering – it is not so clear how semantic and episodic cognition are to be distinguished with regard to the future, however.[63]

62 One might perhaps expect that episodic future thinking and remembering occurred more or less consistently from an introspective, i.e., first-person, point of view, but this does not seem to be the case. Kathleen McDermott and colleagues (2016) found that third-person perspectives predominate both when participants are instructed to remember and to imagine a personal experience. Furthermore, third-person perspectives occur even more frequently in future thinking than in remembering, with more temporally distant future scenarios generating even more third-person perspectives than temporally close future events (McDermott et al. 2016, 246f and D'Argembeau et al. 2011). However, there is growing evidence that individuals typically have a preference for one perspective over the other, in that individuals who are generating more third-person perspective memories in total also generate more simulations of the future from this perspective. Moreover, there seem to be great differences regarding the phenomenology of episodic future thought and memory, affecting, e.g., the vividness of the scenario as well as the associated feelings of presence. Additionally, some individuals engage much more often in episodic future thinking than others (Szpunar 2010).

63 This being said, the specific definitions of semantic and episodic *memory* have been questioned, as well. In light of empirical work on the (lack of) authenticity of episodic memories, so-called 'false memories' and the constructive nature of episodic memories in general, major disagreement arose with regard to the question whether an episodic memory must necessarily involve a causal link to the experience of the event itself, or not. Those defending the constructive episodic simulation hypothesis, primarily psychologists and neuroscientists, consider the causal link unnecessary, whereas others, often philosophers, view this causal link as necessary. For a modern account of this claim, also seeking

Initially, Tulving (1972) argued semantic memory to be about facts and abstract or general knowledge about the world, and episodic memory about personal events including *WhatWhereWhen information,* or *WWW-memory*, that is, what happened (or will happen) where and when. Later, however, it was argued that WWW-memory is neither necessary nor sufficient for episodic cognition. It is not necessary, because one might have an episodic memory of an event, even if one does not remember when or where exactly this event took place, and it is not sufficient, because it seems that one can semantically think about a personal event including its detailed contextual information without necessarily projecting oneself into that specific moment, which has commonly been argued to represent a major part of the dichotomy between semantic and episodic cognition. Tulving concluded that WWW-memory cannot sufficiently distinguish semantic from episodic cognition, therefore. Rather, the difference between both memory systems shifted towards the specific *feel*, that was argued to accompany episodic thinking.

Typically, it is assumed that episodic, but not semantic future thinking involves the individual to engage in some kind of mental projection into the future set-up itself. The ability to do so has been called *mental time travel* (MTT)[64]. MTT can occur forwards and backwards in time in that an individual with mental time travel capabilities can mentally travel into her past and her future. According to Tulving's later writings, a subject with episodic future thinking capabilities necessarily needs to be able to engage in MTT. Mental time travel "takes the form of remembering personally experienced and thought-about events, occasions, and situations that occurred in the past, together with imagining (pre-experiencing) personal happenings in the subjectively felt future" (Tulving 2005, 47). Tulving argues that only episodic memory, but not semantic memory, enables mental time travel, which, in turn, represents the function of the episodic memory system.

As has become clear from the examples above, mental time travel, as defined by Tulving, clearly has phenomenological features, in that it transports the feel and vividness of the remembered or imagined situation and enables the subject to re- or pre-live certain emotions or needs related to the scenario. In his later writings, Tulving changed his initial definition of episodic memory in that it now was said to include a specific kind of consciousness, which he

---

to do justice to the mentioned empirical research, see, e.g., Cheng and Werning (2016). Kourken Michaelian (2017) is one major representative of the constructive episodic simulation hypothesis. His article in the Stanford Encyclopedia of Philosophy provides a good overview of both positions and their (dis-)advantages (Michaelian and Sutton 2017).

64 In chapter 4.3.2, the ability to project oneself into a past or future scenario will be further specified. Furthermore, mental time travel will be distinguished from mental space travel.

termed *autonoetic* (self-knowing) *consciousness* (Tulving 1985). Autonoetic consciousness "mediates an individual's awareness of his or her existence and identity in subjective time extending from the personal past through the present to the personal future" (Tulving 1985, 1). The *autonoetic consciousness approach* emphasizes the specific feel MTT is supposed to have and its ability to bring one right into the scenario itself.

Clearly, the autonoetic consciousness approach poses an extreme challenge to animal researchers. If one wants to investigate whether an animal has the ability for episodic future thinking in contrast to 'mere' semantic future thinking, one allegedly needs to give an answer to the question whether the animal *just knows* about the future event or vividly *imagines* herself being a part of it. Clearly, this is extremely challenging, if not impossible, in the face of nonlinguistic subjects. However, other behavioral measures to test for mental time travel in animals have been suggested.

Tulving himself was sure that the ability, on the one hand, to episodically remember past events and the ability, on the other hand, to imagine personal future events were closely related. Suddendorf and Corballis (2007) refer to this the *Janus hypothesis*. Just like the two-faced Roman god of beginnings, who is simultaneously looking into the past and the future, episodic cognition can occur into the past and the future as well. Episodic cognition, on this view, is defined as *one* cognitive capacity, enabling an individual to mentally travel back and forth in time (for similarities and dissimilarities of MTT into the past and the future, see also Suddendorf 2013a).

In the past decades, Tulving's assumption has been put on stable empirical ground. In a nutshell, there are three strings of evidence arguing in favor of the Janus hypothesis: neurological (e.g., Schacter and Addis 2007, Buckner and Carroll 2007), neuropsychological (for recent overviews, see e.g., Eacott and Easton 2015, and Klein 2016) and developmental (e.g., Suddendorf and Busby 2005). Taken together, these findings suggest that an individual capable of episodically remembering past events is also able to episodically imagine personal future events. On this view, the presence of one ability, therefore, can be taken as evidence for the other one as well.

On the first glance, it might look as if the ability to mentally travel back in time (PMTT) confronts empirical science with the same problem as future mental time travel (FMTT). Yet, Thomas Zentall and colleagues developed a clever experimental design to ask for information the animal did not expect to be asked for (because the question had been part of an earlier experiment). Zentall et al. (2008) hypothesized that when the animal was unexpectedly being asked for a piece of information, which it did not deem relevant at the time the information was present, she necessarily had to mentally travel back in time in order to be able to answer the question. The alleged clue of this *unexpected question paradigm* is, therefore, that the animal can give the right answer only by means of mental time travel.[65]

Despite its elegant design, critics have pointed out that the question could not have been so unexpected after all. As the animals were exposed to extensive training beforehand, they plausibly just constantly held in mind both rules of how to act in situation A and B, so that there was no need to travel back in time at all, when they were unexpectedly exposed to situation A. Thus, the animal could, in theory, acquire the correct answer by different means, for instance, semantic memory (Cheke and Clayton 2010, 919f).[66]

Furthermore, even though a large body of evidence argues in favor of the Janus hypothesis, *discontinuism*, that is, the position that episodic remembering and episodic future thinking are not two instances of one underlying cognitive system, remains at least possible. In investigating animal planning abilities, therefore, I will only consider research directly assessing episodic future thinking in animals and not take empirical research on episodic memories into account.

Researchers have responded differently to the challenges of the autonoetic consciousness approach. Some, such as Zentall et al. (2008), have tried to develop experimental designs that meet its demands, others, such as Corballis (2013), have searched for neurological evidence. Yet others, for instance, Nicola Clayton, have simply oriented themselves on Tulving's earlier definition of episodic cognition (*episodic-like approach*). The basic problem of the autonoetic consciousness approach is that if episodic (future) thinking is strictly defined as to always involve autonoetic consciousness, it may not be empirically accessible. As a result, episodic cognition defined through the autonoetic consciousness approach has been argued to put animal research to a halt.

Consequently, many researchers and theorists alike proposed a change of direction in the research area. Instead of putting emphasis on the phenomenology

65 Related to the unexpected question paradigm are source memory studies. Source memories are a metacognitive type of memory, where participants do not only have to remember the contents of a memory, but also *how* they come to know what they know (Jelbert and Clayton 2017, 104).

66 Claudia Fugazza and colleagues recently presented a modified method of the unexpected question paradigm which might overcome this problem, however. The authors propose that when the animals – here, dogs – were unexpectedly asked to imitate a specific action they had only watched their owners' perform some time ago, but not practiced themselves, they could not rely on motor practice and, therefore, necessarily needed to "recall[…] a mental representation that was formed during encoding" (Fugazza et al. 2016, 3212), i.e., use their episodic memory. This variation of Zentall's paradigm may in fact be a particularly promising way to investigate episodic memory in other animal species as well.

and the mode of consciousness as the defining features of episodic cognition, Eacott and Easton (2012) proposed to pursue a consciousness-neutral definition of episodic cognition, focusing on the empirically accessible behavioral outcomes of future-directed cognition. This strategy may look unconvincing, at first glance, because it allegedly argued that the core feature distinguishing episodic from semantic memory – mental time travel including a specific type of consciousness – *cannot* be essential, simply *because* it is not empirically accessible.

There is more to say about this, however. Recently, with the emergence of the *constructive episodic simulation hypothesis*, an alternative view on the connection between episodic memory and episodic future thinking has gained influence and now represents what can justifiably be called the default position in psychology. The *constructive approach* of episodic cognition acknowledges that both episodic memory and future thinking share certain features, but denies that these can be explained via the concept of mental time travel. Mental time travel does not grasp the essence of episodic cognition, it is argued; rather it is the ability to construct a mental representation of a personal scenario that differs from the currently perceived one. Crucially, it is relatively irrelevant, whether this mentally represented situation lies in the past, the future or is purely imaginative (see, e.g., Eacott and Easton 2012, Hassabis and Maguire 2009 and Schacter and Addis 2007).

Episodic future thinking, on this reading, is primarily about the mental construction of a scene by help of spatial cues, where the temporal context of the scene is only an amendment, but no requirement: “Crucially, this view of episodic memory and future thinking […] removes any need for the concept of MTT or, slightly paradoxically, any absolute requirement for the concept of time in consideration of past and future events” (Eacott and Easton 2012, 205).

To sum up, there is currently strong disagreement with regard to the requirement of a temporal aspect: one side, which has been represented here by Eacott and Easton (2012), pleads to get rid of the temporal constraint of mental time travel and to rather focus on the constructive nature of episodic cognition, while the other side demands a preservation of the mental time travel aspect of episodic cognition.

One consequence of a time-neutral constructive view of episodic cognition is, however, that it might not grasp the kind of developmental and comparative differences with regard to an understanding of the future, and time in general, that have been discussed thus far. For this reason, among others, proponents of the other side partly (re-)turn to Tulving’s definition, while simultaneously acknowledging the plea of empirical researchers not to define episodic cognition in terms of vague introspective and phenomenological features, but to rather clearly conceptualize and explicate the kind of temporal understanding, which they argue to be a crucial component of episodic (future) thinking.

For the purpose of this thesis, it does not play a significant role whether the dichotomy between semantic and episodic cognition is better captured in the first or the second position. As has already been indicated above, the debate has been going on for several decades now and though I am confident that considerable progress is being made, an answer to this problem is not essential for my thesis. More significant for me is that there *are* different ways to anticipate the future which significantly influence the way an individual is capable of thinking about the future. In fact, there may be a way to accommodate both positions to a greater or lesser extent, which will be presented below.

That episodic future thinking crucially involves the simulation of a specific future event is, in fact, broadly shared in the debate around prospective cognition in animals by both sides of the debate.[67] Some argue, however, that not *any kind* of simulation will suffice as evidence, because episodic simulation must be further distinguished from semantic simulation. In specifying this difference, it will become evident, once more, why one side of the debate puts so much emphasis on the temporal aspect of episodic future thinking.

4.3.2 *Semantic vs. Episodic Simulation*

Simulation describes "the construction of a detailed mental representation of a [...] future event" (Szpunar et al. 2014, 18416). The two verbs *simulate* and *imagine* will be used interchangeably here and hereafter, even if simulation, strictly speaking, is not identical with imagination, because simulations, but not imaginations, refer to realistic scenarios only and take certain present confinements into account. If I envisage spending the next four weeks in Hawaii, for instance, even though I have to work on my thesis, then I engage in imaginative daydreaming, but not simulation (Atance and O'Neill 2001, Szpunar 2010). However, the distinction between plausible and fantastic scenarios is one of degree and may, further, turn out to be insignificant in behavioral or neural terms, so it will be disregarded here and hereafter (Szpunar 2010, 145).

67 In fact, the terms *episodic future thinking* and *episodic simulation* are now commonly used interchangeably, even though the distinction between semantic and episodic theoretically applies to other modes of future thinking as well (for example, following Szpunar et al.'s (2014) taxonomy, intention, prediction and planning). A possible reason for this use of words could be that if the term episodic is taken seriously, then an episodic future thought must necessarily refer to an episode, that is, a sequence of events. Thinking about a future episode, in turn, usually occurs by ways of simulation. I take 'event' in Szpunar et al.'s definition to refer to a scenario, rather than a static scene, thus, which is also consistent with the authors' examples. Note, however, that there will very probably be other instances of episodic future thinking besides simulation.

According to Szpunar et al. (2014), the distinction between semantic and episodic simulation is as follows: if the mental representation that is being constructed refers to specific, autobiographical[68] future events, the individual is commonly said to engage in episodic simulation; if it refers to a more general or abstract state in the world, the simulation is said to be semantic (Szpunar et al. 2014, 18416). Of course, not all instances of simulation, and, indeed, future thinking in general, can be strictly classified as semantic or episodic in the above sense, so that Szpunar et al. take the distinction to be on a continuum, allowing for hybrid forms. An example for such a hybrid form would be the simulation of an autobiographical, but not specific future episode. John, for example, could imagine himself becoming a professional cook, without necessarily projecting himself into a *specific* scenario of cooking a meal.

Importantly, and somewhat paradoxically, however, a thus defined episodic simulation does not necessarily involve any kind of mental self-projection, because our semantic memory already provides us with a rich context.[69] Szpunar (2010) illustrates this with the help of Angie, who is preparing to attend her company's Christmas party, which has been taking place at the same venue for the last twenty years. Angie imagines herself entering the banquet hall, wondering what it will look like.

If Angie has just started to work for the firm, she likely depends on an abstract knowledge of what Christmas parties typically look like (*'There will be a Christmas tree and white tablecloths, the decorations will be green, red and gold and there will be a stage for the band …'*). Additionally, she may recall information of another Christmas party she attended in the past. If Angie, in contrast, is in her second year at this company and attended last year's Christmas party, she may recall a very specific incidence related to this event; having a conversation with her co-worker while entering the venue, for example. If, as yet another alternative, Angie has been an employee of this firm for two decades, she will probably not need to remember a specific event, but has acquired an

68 Autobiographical events will be used as being synonymous with personal events.

69 In fact, the role and significance of semantic memory for future thinking has long been extremely underestimated. Semantic memory seems to be particularly important for future thinking; thus, it has been shown that patients with semantic dementia can still remember past experiences, i.e., they still have the ability to recall episodic memories, but have great problems to imagine future (episodic) experiences (Schacter et al. 2017, 42), which has led to the *semantic scaffolding hypothesis*. This hypothesis argues that semantic memory provides the scaffold of any kind of future thought and that episodic details are added only afterwards. If this is true, then semantic memory is necessary for both semantic and episodic future thinking. Osvath and Matin-Ordas (2014) propose, in a similar fashion, that the extraordinary prospective thinking skills of humans are not due to their unique mental-time-travel ability, but a result of their extraordinary semantic knowledge of the world.

abstract representation of how the banquet hall typically looks like through the years (Szpunar 2010, 156). Presupposed that Angie in all three cases simulates a specific, personal future event, that is, the Christmas party, she would be said to engage in episodic simulation in all three cases, according to Szpunar et al.'s (2014) definition, even though she does not always need to engage in mental time travel to do so. If it is acknowledged, however, that there is an important difference between a simulation based on one's semantic knowledge of the world and a simulation based on an episodic memory, then the above examples of Angie's simulations should probably be kept separated and not all be put together under the heading of an episodic simulation.

In order to avoid the somewhat paradoxical implication that an episodic simulation might refer to a simulation that does not rely on the episodic memory system at all, the concept of episodic simulation will be tied closer to its pre-experiential nature. This will lay emphasis on Angie projecting herself into the scenario itself, imagining to step into the banquet hall, greeting co-workers and her boss, watching people chat at the ostentatiously decorated tables and finally searching for a free chair close to the stage. Because only a dynamic sequence of events can be 'pre-experienced'[70] in this sense, a simulation must necessarily occur into a scenario rather than a static scene. True, Angie may have a fairly detailed mental representation of the event's venue in all three cases, but only those instances that involve her to mentally project herself into the scenario will be termed episodic, on this reading.

Self-projection can be understood in at least two slightly different ways, however, which allow for two possible ways to define episodic simulation in contrast to semantic simulation, in turn. For one thing, self-projection could label the feeling of presence when a scenario is to be mentally constructed – rather than just *knowing* what will be the case in the future, the individual projects herself into a scenario, *pre-experiencing* it (Atance and O'Neill 2001).

Importantly, such a representation of a future event could still be "uncontextualized" (Redshaw 2014, 3) in the sense that an individual vividly imagines herself to be in a specific situation, without simultaneously being able to represent the temporal-causal nature of the relationship between this

70 Of course, one cannot literally *pre-experience* an event that has not happened yet. Pre-experience here is taken to refer to the simulation of a future event relying on the mental construction of a scenario, composed of many different elements from memory, including specific past personal experiences (episodic memories), but also abstract or general information (semantic memories) (see also Eacott and Easton 2012). Whenever such a simulation involves one or more episodic details, which may stem from the retrieval of disparate memories of past experiences, the simulation is also described as resulting in a feeling of (pre-)experience in the individual.

representation and current reality. Thus, an individual could engage in what has been called *mental space travel* (MST) in contrast to those individuals capable of mental time travel (MTT) (Redshaw 2014). Just like mental time travel, mental space travel is described as being possible into an individual's past and future. It should be noted that the term *mental space travel* is actually a bit misleading, because mental space travel involves a temporal dimension as well insofar that it refers to the simulation of a temporally extending scenario rather than a static setting. Mental space travel, however, only allows the individual to mentally re- and pre-experience specific scenarios, but not to embed them into a temporal framework.

Under this reading, Jane, from the example above, engages in semantic simulation, because she just knows that a proper dinner consists of appetizer, main course and dessert and that her friends will expect to be served all three courses tonight. Based on this semantic knowledge, Jane prepares dessert. John, on the other hand, imagines himself to be in the embarrassing situation of not having prepared any dessert for his friends. Maybe he episodically remembers a very similar situation or he 'extracts' the relevant episodic detail (the feeling of slight embarrassment) from a more or less completely different situation (for instance, the uncomfortable feeling of having forgotten an appointment) and transfers it into his imagined scenario, which suffices to make him prepare dessert (Eacott and Easton 2012, 205; Cordonnier et al. 2016).

As Hoerl and McCormack argue, it is not clear whether this understanding of episodic simulation requires any understanding of *time* on the part of the individual, however: "Indeed [...] there is a sense in which the purpose of such simulation is precisely to diminish the relevance of time, by allowing the subject to 'pre-experience' the future outcome already, reducing the subjective temporal distance to it, and forging a tighter connection between one's present self and one's future one" (Hoerl and McCormack 2016, 250).

Alternatively, the ability to project oneself into the future deemed essential for episodic future thinking can be understood as to require a transition from one's current position in time (Hoerl and McCormack 2016). This means that an individual with episodic thinking capabilities must be able to take a different temporal perspective than the one she currently holds and thus must be capable of telling which events will lie in the past from this perspective, and which events still lie in the future. An easy way to understand this is to bring to mind that you can make sense of the sentence *Tomorrow, today will be yesterday*. Furthermore, in simulating a *possible* future scenario and not just imagining any kind of hypothetical scenario, the individual must be able to represent the temporal-causal relationships between two events (see also Szpunar 2010 and Atance and O'Neill 2005).

On this reading, John from the example above engages in episodic future thinking not because he can simulate an embarrassing situation, but because he knows that this mental representation refers to one of many possible future scenarios and because he knows that certain events and actions will have an effect on the course of the future, for instance, that not buying ice cream will probably result in an embarrassing situation later.

Jane, in contrast, could set the dinner table by temporally updating her model of the world by 'running off' a simulation, for instance, imagining herself sitting at the table, eating lasagna with a fork and a knife from a plate (McCormack and Hoerl 2011, 133). This motivates her to get plates, forks and knives and place them on the table. Then Janes imagines sipping wine from a glass, so she gets wine glasses; etc. Jane's mental representation of a future state thus functions as the visualization of a goal-state, which motivates her to act in a certain way. Importantly, however, without temporal-causal reasoning abilities, this remains a relatively inflexible way to prepare for future contingencies.

To sum up, there are at least two different ways to define episodic simulation, which both refer to the requirement of pre-experience and self-projection, that is, via mental space travel or mental time travel. As has repeatedly been mentioned, there is considerable dispute as to whether episodic simulation merely requires the ability to 'pre-experience' a future scenario in terms of mental space travel or whether it additionally necessitates an understanding of the temporal-causal relationship of the involved events on behalf of the individual, which has been defined as mental time travel. Rather than arguing for one of the two sides, it will be suggested that both mental space travel and mental time travel should be regarded as two forms of episodic future thinking. Typically, a subject is *either* capable of mental space travel *or* mental time travel, in that both represent a rather different way to simulate future events. Mental time travelers, however, will also sometimes engage in mental space travel, for instance, in dreaming, whereas mental space travelers will never be capable of mental time travel.

4.3.3 *Temporal Dimensions of Mental Space Travel and Mental Time Travel*

In light of a recent study by Donna Kwan and colleagues (2013)[71], Redshaw hypothesizes that in mammals, the temporal dimension associated with men-

71 Kwan et al. (2013) found that hippocampus-lesioned amnesiacs are severely limited in their ability to construct detailed mental representations of future scenarios (see also

tal space travel (MST) may be powered by the hippocampus, whereas the ability to represent temporal-causal relationships (requiring metarepresentation, according to Redshaw) may be located in the extended human prefrontal cortex. Thus, "[t]he temporal dimension associated with the hippocampus [...] may only exits *within* the mental representation itself. It may not allow the representation as a whole to be embedded in a specific temporal context relative to current reality" (Redshaw 2014, 5, Box 1).

One could acknowledge Redshaw's distinction between mental space travel and mental time travel, but argue that mental time travel is not necessary for episodic simulation, in that individuals without mental time travel capacities could be said to engage in episodic cognition nonetheless. On this view, the ability to engage in metarepresentation could be argued to significantly transform the ability to episodically simulate the future, so that episodic simulation in (human) persons through mental time travel may be extremely different from episodic simulation in animals (and small children) through mental space travel. Redshaw also argues, however, that human adults frequently engage in mental space travel, without simultaneously representing the representation's relationship with current reality, so that metarepresentation "is an additional, rather than encapsulated ingredient" (Redshaw 2014, 3) of human episodic simulation. Thus, mental time travel crucially enhances episodic future thinking, but only mental space travel may be necessary for episodic thinking.

I will acknowledge Redshaw's conceptual separation between mental space travel and mental time travel, but remain neutral with regard to the question whether some animals may be able to engage in mental time travel as well and stay silent with regard to the neural locus of this capacity. Furthermore, rather than either arguing that episodic future thinking requires the ability to

---

Hassabi et al. 2007), but still grasp the temporal relation that persists between representations of the future and the present. These patients could be interpreted to have severe limitations in mental space travel, due to lesions to their hippocampus, but to nonetheless have the preserved ability of metarepresentation and, thus, to understand temporal-causal relationships, associated with the prefrontal cortex. A consequence of this hypothesis would be that an individual may be able to form metarepresentations, yet not engage in episodic thinking, or only in an extremely diminished sense. Interestingly, this may be compatible with other findings suggesting that hippocampus-lesioned patients can solve false-belief tasks (associated with metarepresentation) and also seem to have a preserved understanding of the *concept* of anticipated regret, while being unable to *engage* in regret themselves (Redshaw 2014, 5, Box 1, Hoerl and McCormack 2016, 253). Thus, Craver et al. (2014) report that even though patient K.C. apparently knew what (anticipated) regret was, he could not give any concrete examples for when he or another person might feel regret. Taken together, these findings could be interpreted to mean that that hippocampus-lesioned amnesiacs seem to be severely limited in simulating a specific, detailed personal scenario, while simultaneously having a preserved ability to embed these representations into a temporal context and to form metarepresentations.

project oneself into the future by way of mental space travel or by mental time travel, I will simply indicate the type of episodic simulation by either using the term *episodic mental space travel simulation* or *episodic mental time travel simulation*.[72]

By help of Szpunar's Angie-examples, it has been seen that a detailed mental representation of a personal, specific scenario can be constructed in the absence of recalling an episodic detail from a past experience. This simulation has been argued, contra Szpunar (2010), as not being episodic, because it neither involves mental space travel nor mental time travel. Neither, however, is it distinctly semantic. It seems plausible to assume that these hybrid forms of simulation are very common, so that not every instance of a specific, autobiographical simulation is genuinely episodic.

Yet, we are interested in what cognitive preconditions must be given in order to ascribe an individual with the ability to engage in episodic simulation to begin with. Depending on which of the two ways of episodic simulation described above are laid focus on, one will either consider mental space travel, that is, the ability to construct mental representations of scenarios that differ from current reality, or mental time travel, that is, the ability to represent the temporal-causal relationship between two representations, as significant. In sum, rather than arguing for or against the position that episodic future thinking required mental time travel, I will merely conceptually separate two ways to engage in the simulation of a future scenario. Because there is considerable dispute as to whether episodic future thinking merely requires mental space travel or mental time travel capacities on behalf of the individual, both ways to simulate the future will be termed episodic.

Finally, the conceptual distinctions made above can be linked to temporal-causal reasoning and temporal updating. It should have become clear by now that temporal-causal reasoning is essential for mental time travel, but not mental space travel. Furthermore, an individual who can temporally update

---

72 Again, this may be compatible with Szpunar et al.'s (2014) taxonomy of future thinking. As already mentioned, Szpunar argues that a simulation must be conceptually dissociated from imagination in that the former only refers to possible future scenarios. This could be taken merely to mean that the simulated scenario must be realistic, which would only require that present constraints must be taken into consideration. But it might also refer to the ability to represent the future as being uncertain: rather than just represent what will realistically be the case, an individual should represent what *could* be the case. Thus, a possible future scenario could be episodically simulated either by help of mental space travel or mental time travel. An individual with mental space travel abilities may very well be able to mentally represent multiple possible, i.e., realistic, future scenarios, but via mental time travel the individual additionally represents these representations as mutually exclusive.

her model of the world may, or may not, be able to engage in mental space travel, because temporal updating might involve mental space travel or be based on semantic knowledge. Thus, a child might either, via episodic mental space travel, represent first placing the hairbrush into the red and then the blue cupboard or just know (semantically) where the hairbrush now is. In a similar fashion, Jane might simulate eating the main course and afterwards dessert or just know that the main course is going to be served.

What remains to be clarified is which fitness-enhancing benefits mental time travel simulation has over mental space travel and semantic simulation and if and how these different ways to simulate the future enhance the degree of psychological connectedness. Szpunar (2010) and Schacter et al. (2017) argue that there are four major ways how episodic simulation can be beneficial, namely in terms of emotional regulation, goal achievement, temporal discounting[73] and prospective memory.

In detail, simulating a positive image of the future can help to reduce emotional tension in stressful situations. Additionally, it can enhance motivation to achieve the desired goal and find solutions to do so. Process simulation, where the individual is focusing on the necessary steps how the desired goal can be attained, is particularly effective here, whereas outcome simulation, that is, a simulation of the desired goal state itself, is especially advantageous in intertemporal far-sighted decision-making. Typically, future rewards are discounted relative to the time the individual has to wait for them. Simulating the outcome, however, increases patience to wait and shifts the choice from consuming the small, immediate reward towards the larger, but temporally distant reward. Furthermore, simulating to implement a specific intended action at a

73 The empirical findings on the role of episodic memory for temporal discounting are not entirely unequivocal, however. Kwan et al. (2013) found, contrary to Boyer's (2008) hypothesis, that amnesia patients unable to generate episodic memories of the past do not tend to discount the future any more than healthy controls, which seems to suggest that episodic memory does not moderate temporal discounting (see also Craver et al. 2014). One interpretation of these findings would be that *some* kind of simulation moderates temporal discounting, but that it is semantic rather than episodic simulation. Alternatively, the ability to form metarepresentations could be argued to moderate temporal discounting, located, according to Redshaw (2014), not in the hippocampus but the expanded human prefrontal cortex, so that lesions to the hippocampus would impair the ability to form episodic memories, without having an effect on temporal discounting. So far, the meaning of the episodic memory system for temporal discounting remains to be unclear. Further research will be needed to disentangle the role of semantic and episodic memory for temporal discounting.

precise time and place raises the probability of actual enactment.[74] Put differently, the simple act of forming a specific intention makes it likelier that the intended action will be executed later. Planning can be regarded as offering a promising avenue to study episodic future thinking, because a good plan very often involves the ability to vividly envisage a future scenario; even more so, if the simulated situation will be new to the subject.

Emotional regulations, outcome simulations and implementation intentions refer more directly to episodic future thinking in terms of mental space travel, because they emphasize visualization, whereas process simulations becomes particularly beneficial in connection with mental time travel. The reason for this is that envisioning the necessary steps to achieve a goal is particularly effective in situations where one tries to find the best among several possible solutions. Preparing for an upcoming test, for instance, typically involves the ability to mentally change the temporal order of events and actions and to compare different scenarios to one another. A good plan also takes certain facts, constraints and fixed dates into account, such as the knowledge that one can learn better during night-time or that there still needs to be some time to do grocery-shopping and to take the dog for a walk. Doing this, however, requires temporal-causal reasoning, which has been argued to be essential for mental time travel.

The most significant fitness-enhancing benefit of a simulation on basis of mental time travel in contrast to mental space travel, however, becomes evident in one particular ability which has already been discussed in the last chapter, namely regret. Hoerl and McCormack (2016) even argue that experiencing and anticipating regret represents the distinctive function of episodic memory for decision-making. What makes regret different from other negative emotions is that it goes beyond a mere reliving of an emotion associated with a certain choice. It brings awareness to the fact that at one point in the past the future had been undecided so that an alternative decision would have led to an alternate reality than the current one.

Thus, regret requires the individual to consider one point in time from two temporal angles, a backward-looking perspective, from which the relevant decision has already been made, and a forward-looking perspective, from which the decision can still be made (Hoerl and McCormack 2016, 248f). Furthermore, it requires the individual to think about points in time independently from the events that did or could have happened then, an ability which the authors term *event-independent thinking* (McCormack 2014). Analogously

74 Implementation intentions are similar, but not identical with process simulation, because they do not necessarily involve any kind of simulation (Szpunar 2010, 155).

to the child who understands that the toy mouse could have emerged from the other exit of the branched tube, an individual capable of feeling regret knows that event B rather than A could have happened at t_1. Anticipated regret refers to the simulation of how someone will feel looking back at a decision which the individual has yet to make, knowing that she will have to live with the consequences of her decision.

Of course, regret does not always involve the retrieval of an episodic memory, because we do not always remember the specific circumstances and context of a past decision. Thus, experiencing regret does not necessarily seem to involve episodic simulation. Nonetheless, the *ability* to experience regret depends on the specific way to represent time underpinning mental time travel. Mental time travel has been defined in terms of a projection of the self into a temporally distinctive perspective, where the individual can represent the temporal relationship between this perspective and the presence. Because of her temporal-causal reasoning abilities, thus, the individual knows which events, choices and so on are already in the past from this new perspective and which ones still lie in the future. To illustrate this with an example by Hoerl and McCormack, you may not be able to episodically remember drunkenly writing an email to your boss in the middle of the night, but you can still mentally travel backwards in time and reconstruct (rather than recollect) what must have happened then and, importantly, regret your decision of which you know that you must have made it at some point, despite being unable to remember doing so (Hoerl and McCormack 2016, 251).

Even though regret often leads to the insight that a past decision has been bad and can thus serve as a reminder to decide better in future similar situations, regret is not necessary for this kind of insight. An individual incapable of regret is still able to learn from her mistakes and avoid these in future similar situations, for instance. Hoerl and McCormack thus suggest that the specific beneficial nature of regret becomes particularly evident with regard to anticipated regret. They argue that anticipated regret may have the function, first and foremost, to remind us that certain choices cannot be undone and that one might have to live with their consequences for the rest of one's life.

4.3.4 *The Relevance of Simulation for Psychological Unity*

Analogously to the last chapter, it will be laid out to what extent the three different ways to simulate the future that have been differentiated above contribute to the strength of prudential unity in terms of McMahan's time-relative interest account.

An animal who can semantically simulate the future may very well be able to anticipate a possible personal event, as has become evident in the Angie example. Furthermore, such an individual can even, to a certain extent, anticipate her own future mental states on the basis of general knowledge or personal traits. Jane, for instance, knows that she will want to serve dessert without having to project herself into the specific scenario itself. Angie can give a fairly detailed description of the banquet hall, because she has attended many similar Christmas parties in the past. Thus, semantic simulation might enable an animal to anticipate her future mental states and, thus, strengthen the psychological relationship to her future self.

Depending on the role one allows semantic memory to play for future thinking, semantic simulation might provide the animal with a more or less sophisticated capacity to simulate the future. The importance of semantic memory for future thinking is still hotly debated, so it might remain unclear as to how much exactly semantic simulation contributes to psychological connectedness and continuity.

What is clear is that episodic mental space travel simulation enables an individual to mentally construct a specific, context-rich future scenario and be a part of it. Future events and associated mental states are vividly experienced. The purpose of mental space travel simulation has thus been described as "to diminish the relevance of time, by allowing the subject to 'pre-experience' the future outcome already, reducing subjective temporal distance to it, and forging a tighter connection between one's present self and one's future one" (Hoerl and McCormack 2016, 250). Perhaps unsurprisingly, then, this type of simulation has been shown to be beneficial in terms of emotional regulation, motivation to pursue a goal, temporal discounting and prospective memory.

This tighter connection between present and future self, that Hoerl and McCormack point towards, should provide us with sufficient reason to think that an animal with the ability to episodically simulate the future in terms of mental space travel is, *ceteris paribus*, psychologically stronger connected to her future self than another subject who can merely semantically simulate the future.

Yet, even though episodic mental space travel simulations already possess a temporal dimension in that the associated mental representation constitutes a dynamic scenario rather than static scenes, this kind of simulation is also atemporal in the sense that the individual cannot put the representation into a temporal context, relative to present reality. This means that, somewhat paradoxically, an animal with mental space travel abilities can mentally represent a short sequence of events and even act in a way as to pursue the mentally represented goal state, without being conscious of the fact that this representation refers to the future.

At first glance, this seems strange, because even individuals with mere mental space travel capabilities do not mistake their representations for reality and also seem to know whether a specific mental representation indicates a goal state, referring to the future, or provides information, acquired in the past. How should they be able to do this in absence of a metarepresentative insight into the temporal-causal relationship between these representations and current reality?

Redshaw suggests that mental space travel may have been naturally selected in that it functions as a mental 'training' of certain adaptive behaviors (see also Boyer 2008). Involuntary mind-wandering during wakefulness, for instance, could cue an individual to actually perform a specific action, when the possibility to do so arises in the future (Redshaw 2014, 5). Similarly, a dreaming rat may succeed in finding the correct path through a maze, so that when she is actually confronted with the maze again, she easily finds her way through. Environmental cuing could, in a similar fashion, trigger adaptive future-oriented behavior. Redshaw takes the example of hungry chimpanzee, who comes across some stones. This fact triggers in the animal's mind a representation of himself using these stones to crack nuts, so that the chimpanzee is motivated to take these tools with him to a site, where nuts are available (ibid. 6).

What these examples show is that mental space travel can result in future-oriented, fitness-enhancing behavior without the individual having to grasp the relative temporal dimension of the representation, for instance, the temporal-causal relationship between the mentally represented scenario and the present. This raises the question how episodic simulation through mental space travel can be empirically distinguished from mental time travel at all. If mental space travel is enough to solve a novel problem the animal has never experienced before, then there must be another way to behaviorally separate it from mental time travel. And here, finally, I come back to what has been laid out at the beginning of chapter 4, namely the ability to represent what has been termed branching futures.

Redshaw proposes to focus on the temporal aspect of mental time travel and to test whether animals grasp that there can be multiple, mutually exclusive options of a future event. If an animal is able to prepare for both alternatives in the Y-shaped tube experiment, for example, then she also seems to understand that her mental representation of the desired object emerging from one of the two exits may be incorrect. Thus, the experiment "would require the animal to represent the uncertain nature of the relationship between current reality and the future" (ibid. 9). In sum, experiments testing for an animal's ability to represent branching futures may be a good and minimal first measure to test for animal's mental time travel capacities.

I have argued that, despite their inability to represent past, present and future in the sense discussed above, individuals with mental space travel abilities can still be said to be psychologically bound to their future selves to a greater extent than animals who can merely simulate the future semantically. From the discussion of Hoerl and McCormack's hypothesis regarding the main function of what has been called episodic mental time travel simulation, that is, regret, it hopefully has become evident why this type of simulation typically further strengthens the degree of psychological connectedness and continuity. The ability to experience and anticipate regret multiplies the direct psychological connections that persist between an individual in the present and the past or the future, as has been seen in chapter 4.2.4.

Moreover, regret has another consequence, namely, that it significantly shapes what McMahan terms psychological unity. Hoerl and McCormack accurately point out that we often value, in ourselves and in others, new experiences and also the personally transformative character of certain choices. Especially young people are being encouraged to strive for new experiences and a major reason for doing so seems to lie in the fact that we aim to minimize future regret. We know that people mainly regret lost opportunities and their hesitation to dare something novel. In contrast, people seldom regret their decisions of trying out something, even if this should prove to be inapt later. A young woman, dreaming of medical studies, who realizes during practical training in a hospital that the actual job profile of a physician does not match her expectations, will probably not come to regret the internship but be glad about the gain in experience; even if this does not preclude other negative emotions such as disappointment, of course.

Let me sum up the results of this chapter. Despite the widespread consensus on the essentially constructive nature of episodic future thinking, there still seem to be at least two different understandings of episodic simulation in the literature. Whereas episodic mental space time simulation emphasizes the experiential character of personal scenario construction, episodic mental time travel simulation focuses much more on a grasp of the temporal dimension, which enables the individual to additionally represent the nature of the relationship between the mentally constructed scenario and its temporal relation to current reality.

The capability to engage in episodic mental time travel simulation has been argued to typically strengthen the degree of psychological unity between the present and the future self in contrast to the other type of episodic simulation, which, in turn, results in a higher degree of psychological unity, *ceteris paribus*, than the ability to engage in semantic simulation. However, the role that semantic memory plays for future thinking in general has been hotly debated recently, so that its importance for psychological unity may be considered greater in the future.

4.4 Towards a Better Understanding of Planning

As has become apparent in chapter 3, McMahan conforms to the traditional stuck in time hypothesis, arguing that animals are incapable of planning, precisely because of their alleged inability to mentally transcend the current reality. Animals, he argues, are unable to anticipate the future, including their future mental states, and are, in this sense, mentally stuck in the present.

Chapter 4.2. showed, however, that there are several ways how the future can be anticipated, resulting in different ways how a mental transition from the present may occur. First, the future can actually be predicted rather differently, depending on, among other things, the ability to understand that there may be several possibilities how the future will unfold. This also affects an individual's ability to anticipate her (possible) future mental states, as has been seen.

Furthermore, three ways how to simulate the future have been conceptually separated from one another in chapter 4.3. Whereas semantic simulation enables an individual to anticipate a future mental state based on general facts and abstract knowledge, episodic simulation involves a mental projection of the self into the future. Because this self-projection, however, can either refer to the feeling of presence in the to be imagined future scenario or to the ability to consider a temporally distinctive perspective additionally to the present one, episodic simulation, as understood here, can either occur via mental space travel or mental time travel.

As has already been mentioned, planning involves other modes of future thinking as well, which means that the way how an individual sees the future – how she predicts and simulates it, for instance – affects her ability to plan as well. Planning, in turn, has been argued to be particularly well-suited for the purpose of this thesis, because planning typically has a behavioral output, which makes it easily empirically accessible (in contrast to prediction and simulation, for instance), and because planning represents a crucial component of McMahan's understanding of strong psychological unity in that an individual capable of planning should have the potential to form strong psychological bounds to her future.

One reason for why the term *planning* currently looms large in the current debate around prospective cognition in animals is that it has partly replaced the more ambiguous term *mental time travel*. This being said, people still continue to disagree on the definition of planning. Consequently, two sides in a debate might acknowledge that a certain individual anticipates a future (personal) situation and behaves in a future-oriented manner, and still disagree on whether this behavior should be termed planning or not.

It will be argued that to a significant extent, the underlying conflict results from a discrepancy between the common use of the concept planning in daily language and a stricter planning concept, arguing that planning presupposes a grasp of temporal-causal relationships. The latter understanding of planning, thus, is closely related to mental time travel. Whereas some, most notably animal cognitivists such as Mathias Osvath and Nicola Clayton, aim to define planning rather broadly, others, among them Christoph Hoerl and Teresa McCormack, defend a more narrow definition. The debate around Santino, the now-famous stone-throwing Swedish chimpanzee, exemplifies the underlying conflict.

4.4.1 *Defining Planning in a Comparative Approach*

One example of an allegedly planning animal that made headlines is Santino, a chimpanzee living in a zoo in Furuvik, Sweden. Santino became notorious for his attacks on visitors standing outside the enclosure. Hiding in a blind on five consecutive days, a zoo keeper observed that Santino calmly gathered stones[75] in the morning, hours before the zoo opened, and stored them in several caches. Santino even further improved his attacks during the next season in that he hid his caches under concealments, unobservable to the visitors. The chimp used both natural hiding places (for example, logs) and artificial concealments from hay, which he brought from the inside enclosure. He also inhibited aggressive displays before and during the attacks and even engaged in deceptive behavior (slowly strolling towards the caches, fishing out an apple from a water pound and taking a bite), seemingly to create a neutral image. Both behaviors enabled him to throw stones at a much higher frequency from a closer range at the intruders (Osvath 2009, Osvath and Karvonen 2012).

Besides concerns about what this sort of aggressive behavior revealed with regard to the ethics of keeping primates in zoo, the question soon emerged whether Santino could be rightfully described as to *plan* his attacks, or not.

To Mathias Osvath, who first reported Santino's behavior in 2009, the chimp clearly planned ahead in time when he collected and stored stones that he later used as projectiles. His interpretation, however, has been criticized from various sides, for example, by Suddendorf et al. (2009), Suddendorf (2013b),

75 The following year, Santino additionally stored and used pieces of concrete, which he broke off bigger rocks in the enclosure.

Shettleworth (2010a), Roberts and Feeney (2010) and also McCormack and Hoerl (2011), suggesting that Santino's behavior could be explained otherwise, without having to ascribe the chimpanzee with any underlying planning capabilities. As Santino displayed his behavior spontaneously rather than in an experimental set-up, alternative explanations could not simply be ruled out by control experiments. Thus, the Santino debate uncovered an interesting methodological discrepancy within the area of animal cognition, because it revealed different positions on what should rationally be regarded as more parsimonious or simpler explanation in general.

Roughly speaking, one side holds that if there are alternative explanations for Santino's behavior that do not have to assume any planning abilities, these more parsimonious explanations are to be preferred over Osvath's interpretation. The other side argues that ascribing the chimpanzee Santino, a very closely related animal to us human beings, with the ability to plan, should in fact be regarded as much more parsimonious than to assume that even though he behaved *as if* he planned ahead, he did in fact not. Accordingly, it seemed that the Santino debate could only be resolved by reference to methodological principles. I will argue, however, that the Santino debate points to a much deeper and more covert dissent within the research area of comparative cognition, namely that between top-down and bottom-up approaches.

4.4.1.1 On the Epistemic Value of Simplicity

To this day, the epistemic value of simplicity or parsimony[76] is widely acknowledged among empirical scientists working in the research field of animal cognition. Dispute, however, typically arises regarding *which* explanation represents the more parsimonious one, which is, at least partly, due to the ambiguity of the concept itself. An interpretation can be simpler with regard to numerous aspects (for a concise overview, see Fitzpatrick 2009). One of the most famous principles of simplicity is *Morgan's Canon* (Morgan 1903), which argues that one should not ascribe higher mental faculties to an individual if the behavior in question can also be explained by postulating lower mental faculties. Morgan's Canon, thus, would favor a non-planning interpretation over rival explanations in Santino's case.

---

76 Both terms will be used interchangeably. The discussion about the epistemic value of principles of simplicity and parsimony has become increasingly complex, so the following remarks will inevitably not suffice to do this topic justice. As will become evident, however, I will argue that references to (evolutionary) parsimony might not so much express strong claims with regard to the epistemic value of simplicity anyway, but might rather hint towards an underlying plea to bind the concept of planning not too closely to its appearance in human persons.

The epistemic value of simplicity has been highly criticized, however. Jerry Fodor ironically postulated – in allusion to Morgan's Can(n)on – *Fodor's Pop Gun*, which proposes that "in no case may we interpret an action as the outcome of the exercise of a lower psychical faculty if it can be interpreted as the outcome of the exercise of one which stands higher in the psychological scale" (Fodor 1999, 4). Confronted with Fodor's Pop Gun and Morgan's Canon, Fodor then goes on:

> I would say that the two methodological imperatives are both, and equally, silly. A priori, there is nothing to choose between the hypothesis that a certain behaviour has a 'higher' sort of mental aetiology and the hypothesis that it doesn't. If you want to know which is true, you just have to go and look. I am therefore interested in why, while Morgan's Canon is universally treated with great solemnity, [Fodor's Pop Gun] is merely sniggered at. (ibid.)

Similarly, Cecilia Heyes called unfruitful debates about what explanation was the more parsimonious one a "procrastination" (Heyes 1998, cited in Fitzpatrick 2009, 268)[77] and suggested the following: if the empirical data is not unambiguous and allows for two or more interpretations, then more experiments are needed to rule out rival explanations.

Simon Fitzpatrick (2009) agrees with Fodor and Heyes insofar that a simpler explanation (where simpler is understood here in terms of Morgan's Canon) in itself is not a virtue, but also claims that the right reference to parsimony can be justified. We do not have to empirically test for all possible interpretations there are, but can rule out some explanations right away. If one theory A has to make certain assumptions that conflate with, for example, what is known about the species the animal in question is a member of, and the alternative theory B does not make these assumptions, but is in accordance with what is known about the specific animal species, then B is methodologically in favor over A.

Finding a consensus on *what exactly is known* about a particular species may not be so easy with regard to animal planning abilities, however, because it seems that there is actually little on which proponents and opponents can agree on here. Yet, it might be argued to read the deflationary account of simplicity in the context of evolutionary theory[78], because it provides an informative and widely accepted background theory about what is known about the

77 Heyes did so in the context of the mindreading debate.

78 Frans de Waal even suggests to rename the whole research field, commonly known as animal cognition or comparative cognition, into *evolutionary cognition,* arguing that only from an evolutionary standpoint we were able to understand and appraise animal intelligence (de Waal 2016, 28).

phylogenetic relations between two or more species. Accordingly, it has been proposed to understand simplicity in terms of evolutionary parsimony.

Evolutionary parsimony, as understood here, states that if two closely related species show a similar behavioral outcome, then it is evolutionarily or phylogenetically more parsimonious to postulate that this behavior is based on homologous cognitive mechanisms, inherited from a common ancestor. Put differently, if two closely related species, for instance, human beings and chimpanzees, exhibit behavior B and this behavior is assumed to rely on mental state M in case of one species, for instance, human beings, then it is more parsimonious to posit M in terms of the other species, that is, chimpanzees, as well (Fitzpatrick 2009, Sober 2012). It would be simpler, on this approach, to ascribe human-like future-thinking to Santino's projectile-caching behavior than to propose two different cognitive mechanisms in humans and chimpanzees (see, e.g., de Waal and Ferrari 2010, 202; de Waal 2016, 269; Osvath and Martin-Ordas 2014, 6; Corballis 2013).

It is perhaps unsurprising that many cognitive ethologists and comparative cognitivists with a biological background tend to emphasize phylogenetic parsimony. Frans de Waal even proposes to regard mental continuity as the default position in comparative cognition, that is, to presuppose similar cognitive mechanisms "for at least all mammals, and perhaps also birds and other vertebrates" (de Waal 2016, 269), until evidence suggests otherwise. As a consequence, philosophical speculations about what might set humans apart from all other kinds of animals are often regarded with a certain impatience by this side.

And indeed, in light of Darwin's famous words that the difference between humans and all other animals is merely one of degree rather than kind, it seems almost blasphemous to deny mental continuity between all animals, including the human animal (Darwin 1871/2004). In fact, it may very well be the case that the search for a single exclusively human psychological capacity will not be fruitful (Osvath and Martin-Ordas 2014, 7; de Waal 2016, 93). On the other hand, however, it seems equally absurd to deny that humans and chimpanzees are also very different from one another. Especially *because* we are so closely related, it becomes an extremely interesting question *why* we are yet so dissimilar. A strong understanding of phylogenetic parsimony, as expressed, for example, by de Waal, is in constant danger of overemphasizing our similarities and disregarding our dissimilarities.[79]

79 Crucially, this is not to argue that phylogenetic relatedness has no epistemic value whatsoever. From the fact that two species show behavior B and this behavior is assumed to rely on mental state M in one species, it does not follow that there is *strong evidence* for M

I also believe, however, that this rejection of a strong phylogenetic parsimony does not affect as many as one may think in the discussion around animal planning abilities. Nearly everyone would agree that alternative non-planning explanations cannot be ruled out *merely* by recourse to the close genetic relatedness between two species. If a certain behavior could either indicate planning abilities or represent an associatively learned behavior, for example, then the experimenter will naturally strive to rule out one or the other explanation in subsequent experiments. Thus, the tendency to respond to skeptics, who might have raised doubts against the animal planning explanation, by appeals to phylogenetic parsimony, might not so much express a strongly held belief regarding the epistemic value of simplicity, but rather reveal a deeper dissent within the research area of comparative cognition between top-down and bottom-up approaches of how a concept (in this case, planning) should be defined at all and what it exactly entails.

Defenders of phylogenetic parsimony commonly neither intend to argue that control experiments, with the purpose to rule out alternative interpretations (associatively learned behavior, for instance), are superfluous, nor do they deny that a less complex cognitive mechanism could explain the animal's (or human's!) allegedly planful behavior. Rather, they seem to take offense at their opponents' tendency to *acknowledge* their empirical results but *deny* that it represents planning abilities. Osvath, for example, does not argue, of course, that Santino *must* be planning merely on the basis of the close relatedness between chimpanzees and humans. Rather he pleads not to rule out the mere *possibility* of a planning chimpanzee right away by defining planning extremely anthropocentric. Thus, to accept that chimpanzees and human beings may be able to plan for their future is not to argue that they do so in exactly the same kind of way. This leaves room for dissimilarities. For instance, Mathias Osvath and Gema Martin-Ordas state:

> By the logics of evolution, which surely apply to cognition, closely related species sharing complex features not found to the same extent in any close out-group do so because of homologies. It violates parsimony to assume radically different cognitive mechanisms behind highly similar behaviours. If the similarities are accepted as homologies, this opens up the possibility for detailed investigations into the differences, which could then be used for the better to define and explain future-oriented cognition both in humans and other species of great apes in a more grounded way. (Osvath and Martin-Ordas 2014, 6)

---

or that the existence of M is *probable* in the other species as well, however, even if it can be read as probability-raising (Sober 2012). On this understanding, phylogenetic relatedness gains its epistemic value not so much in terms of parsimony, but rather in terms of statistical probability. As I would like to point to another problem, I will not delve into Sober's (and others') notable work on this issue here.

4.4.1.2 Neither Anthropofabulation nor Anthropomorphism

It will be assumed that a methodological approach is *top-down* if it defines a certain cognitive capability p in anthropocentric terms, that is, according to how p manifests itself in normal human adults, and then searches for p in other subjects, such as children, amnesic patients and non-human animals. This is similar to what Buckner (2013) calls *semantic anthropocentrism. Bottom-up approaches*, in contrast, start with a much broader concept of p. P is defined by one or several essential features so that it can be clearly distinguished from other capacities. Instead of searching for human capacities in other animals, a bottom-up approach acknowledges that there might be multiple manifestations of a certain cognitive capability (see also de Waal and Ferrari 2010, Osvath and Martin-Ordas 2014).

The debate around foresight and planning abilities in animals, with its focus on subjectively felt phenomena surrounding mental time travel, has clearly moved in the framework of a top-down approach for a long time. Suddendorf, for instance, argues from a top-down perspective, stressing the uniqueness of human mental time travel and focusing on questions of how and why it emerged. Similarly, the settlement for the term *episodic-like cognition* in case of animals involuntarily re-emphasized the allegedly special status of the human capacity, because it considered 'real' episodic cognition to be exclusively human. Top-down approaches, however, run the risk of defending a tautology. As Mathias Osvath and Gema Martin-Ordas put it:

> It is curious that so many try to define future-oriented cognition and cognitive foresight as close as possible to the human model, such that failure to meet even one part of the criteria means that the animal is not foresighted; taken to the logical extreme, this results in the empty conclusion that only humans have human skills. (Osvath and Martin-Ordas 2014, 7)

Top-down approaches typically have an interest in emphasizing the dissimilarities between two groups of individuals. With regard to developmental or neuropsychological research, this strategy is promising, because here research is interested in the achievement or the loss of p, as defined by its typical appearance in paradigmatic human adults. Applying this approach to the field of comparative cognition, however, is of limited use, because it tends to downgrade animal capacities and simultaneously upgrade human capacities. (It should be emphasized that this coupled downgrading and upgrading is not necessarily implied in a top-down approach, but represents a risk.)

As human beings, we often idealize our own capacities, which has been called *confabulation*. Much more often than we might want to believe, our behavior results from simple unconscious mechanisms, rather than complex reasoning. As a result, the definition of a certain capability in a top-down approach might be too demanding even with regard to humans. Thus, the top-down approach often – even though not necessarily – commits two errors: semantic anthropocentrism and confabulation. Cameron Buckner entitles a combination of the two as the error of *anthropofabulation* (Buckner 2013, 861).

Additionally, testing conditions for children and animals often highly differ from one another, raising doubts as to whether the findings can and should be compared at all (for a good overview, see de Waal 2016). Whereas children, for example, are generally tested while sitting on their parent's lab, chimpanzees are kept behind bars. Crucially, this means that regarding children, the so-called Clever Hans effect cannot always be ruled out, because parents (or experimenters) may involuntarily and non-verbally communicate the correct answers to the children but not the animals.

Moreover, animals are often tested in tasks designed for human requirements, meaning that the animal's specific *umwelt* might not be considered. Buckner (2013) calls this *methodological anthropocentrism*. Human partners in cooperative or other social cognitive tasks typically are of the same species for children but not for the animals, even though it has been shown that animals often fare badly when being tested with humans, but fare equally well as the children, when allowed to interact with conspecifics (e.g., Prétôt et al. 2016). In a similar fashion, many animal species, not so much relying on visual but, for instance, olfactory cues, were long considered to be incapable of self-recognition, because they repeatedly failed the mirror test. When the experimental set-up better matched the specific individual's *umwelt*, however, some animals could be shown to recognize themselves after all.

A project like this one that strives to compare human and animals future thinking and planning abilities to one another, must be careful not to commit the error of anthropofabulation. In other words, one needs to define planning in such a way as to clearly distinguish it from other capabilities, but do not tie it exclusively to its special appearance in humans. A bottom-up approach helps to avoid anthropofabulation. Before it will be laid out in more detail, I will address another error that should be avoided.

It is true that for a very long time philosophers, religious scholars and psychologists have made great efforts to artificially isolate the human from the animal kingdom. In recent years, however, this trend can be argued to have reversed into a "tendency [...] to emphasize the human-like in animals" (Shettleworth 2010a, 479). Sara Shettleworth makes the following observation:

> [A]lthough the extent of human-animal cognitive similarity is undoubtedly a key issue for comparative psychology, it sometimes seems the agenda is to support anthropomorphic interpretations rather than to pit them experimentally against well-defined alternatives. The enthusiasm of the popular and even

> scientific press for clever animal stories nourishes this tendency: killjoy explanations are less likely to make headlines than stories about how octopi or birds are unexpectedly human-like. (Shettleworth 2010a, 478)

Thomas Suddendorf similarly warns that "we need to consider that a persistent flow of published positive findings and a lack of acknowledged negative results may seriously skew our perception of animals capacities" (Suddendorf 2013b, 60). Avoiding both anthropofabulation and anthropomorphism thus demands to take a neutral, curious perspective on animal and human mental capabilities alike. At times, this will imply to dampen our "enthusiasm [...] for clever animal stories" (Shettleworth 2010a, 478), at other times it will caution us to question our alleged superior status. Rather than argue whether or not the difference between animals and humans might be better captured as a gap or as a matter of degree, a comparative project should attempt to capture both the similarities and the dissimilarities.

Buckner agrees in stating: "Moving beyond oversimplified 'pro-' or 'anti-' animal thinking, a healthy comparative project must avoid both biases – by electing fair criteria for experiments to assess, and then relying on objective, empirical assessment of whether those criteria have been satisfied" (Buckner 2013, 868).

4.4.1.3 Towards a Minimal, Bottom-up Definition of Planning

> [W]hat about Darwin's differences in kind and in degree between humans and others? What about the rope of phylogenetic continuity with its two ends? How do the various claims for human uniqueness [...] fit into the picture? There is no profound mystery here, and no need to create one. There are things that bees, birds, and humans all do. But there are also things that bees do but birds and humans do not. There are things that birds do that bees and birds [sic] do not. And there are also things that humans do but bees and birds do not. All living creatures are similar to one another in many ways, and they are also different from one another in many ways. The differences are neither in kind nor in degree; they are differences in light of similarities. Life is part of an endless matrix of sameness and diversity that is nature. (Tulving 2005, 49)

In order to avoid both anthropofabulation as well as anthropomorphism, Buckner (2013) sketches a strategy as to how proceed in defining an ability in a comparative approach. I will apply it to the relevant research question, that is, whether (some) animals can be ascribed with the ability to plan.

The first step is to find a consensual and unambiguous definition of the concept. This means that the definition should be as uncontroversial as possible and that from the explanantia alone one must be able to understand the meaning of the concept. Only after planning has been unambiguously and uncontroversially defined, the general competence criteria, that are embedded within the concept, can be further defined. As has been argued above, it must be avoided to determine these competence criteria along of what will be called a *pinnacle definition*, that is, a definition that unnecessarily demands certain criteria that are met only by the most exceptional planning performances, exceeding other, maybe much more typical, occurrences. Furthermore, uniquely human competences (verbal reports, for instance) should be excluded from the definition as far as possible. These two steps serve the purpose to arrive at an ability's *minimal definition.* This definition then paves the way for the definition of operational criteria, which serve the function to verify planning skills in a specific experiment.

Importantly, Buckner developed this approach with the aim to specify especially vague concepts broadly used in the comparative cognitive literature. It does not imply that each and every capability should be defined according to this principle. It demands us, however, to start bottom-up rather than to define a term according to the performance of a small group of individuals right from the beginning on, and then search for the exact same skills in other individuals. This is similar to Osvath and Martin-Ordas' admonition not to attempt to find evidence for the tautology that only humans have human skills. A thus understood bottom-up approach sets the uncontroversial baseline for a planning concept, and, from here, encourages further conceptual specification of, for instance, chimpanzees', scrub jays' or infants' planning capabilities.

4.4.1.4 Santino Revisited

Above, it was argued that a default position of mental continuity, similar to the one that de Waal seems to defend, is of no real help in deciding whether Santino was planning or not. Because Santino's behavior was not observed in a controlled experimental environment, associative learning (and other) explanations cannot be completely ruled out, even if at least associative learning might be argued to be rather unlikely, because one would have to come up with a plausible explanation of how the association was acquired, in the first place (Osvath 2009, 2010; Osvath and Karvonen 2012). Associatively learned behaviors can be quite elaborate, but they are not very flexible. The fact that Santino neither built haystacks behind other existing objects in the compound, nor manufactured caches in the indoor enclosure or in other, unsuitable places, make an associative learning explanation extremely unlikely. One would expect Santino to make many more 'errors', if his behavior were the result of associative learning.

Importantly, however, this does not yet allow the inference that Santino actually *planned* his attacks, because there may be additional alternative explanations available that do not require planning capabilities. Planning still needs to be defined more precisely.

4.4.2 *Dissociating Anticipatory Planning from Strategic Planning*

Above, at the beginning of chapter 4, planning has been defined as "the identification and organization of steps needed to arrive at a specific autobiographical future event or outcome" (Szpunar et al. 2014, 18418). In the following paragraphs, I will attempt to further clarify what planning essentially refers to in contrast to other intentional acts and conceptually separate anticipatory planning and strategic planning.

Planning has lots of different meanings in folk psychology[80]. By saying that someone *made a plan* or *has planned something out*, we typically mean that this subject has (given herself) a certain goal she wants to achieve, which is located in the more or less distant future and involves several sub-goals. To be considered as a plan, however, these sub-goals need to be brought into a specific temporal order. A shopping list, for example, is commonly not considered as a plan, because the sequential order, in which the items on the list are purchased, is not taken into account. Similarly, not any project will be regarded as an instance of planning, even if the ultimate goal might lie in the very far future and involve quite a lot of tasks. In contrast to a to-do list, which will be defined here as a mere accumulation of tasks that do not stand in any temporal relation to one another or other events, a plan necessarily seems to refer to sequences of events.

To make this clearer, take the plan of achieving a PhD. Assume that this is a desire you presently hold, that is, for which you have considerable present motivation. As a PhD is a rather complex achievement, however, and stretches over a significant period, one needs to break it into many smaller work packages, containing numerous to-dos, in turn. For some of these to-dos it may be

80 Some may get the impression that in analyzing planning by recourse to its manifestation in human persons, we commit the error of anthropocentrism once again and arrive at a pinnacle definition, which has been argued against in the last chapter. This objection can be refuted, however. Semantic anthropocentrism is the inevitable starting point, but sensitivity to anthropofabulation and the aim to define planning as minimal as possible should provide a consensual and unambiguous definition of planning in the end.

relatively, though not completely, irrelevant at what specific point in time they are being carried out. To hand in your thesis, you will have to submit a signed *affidavit*, but at what time exactly you sign and attach it to the thesis is irrelevant, as long as it is included in the final document.

Most tasks, however, are very much dependent on the execution of other tasks, specific dates, environments, other persons and so on. You will only be able to present a paper at a conference next month, for example, if you have finished and analyzed a relevant experiment by then, because its findings constitute a major part of the paper. A successful planner is not only aware of all the required steps to achieve her goal, thus, but also needs to be aware of the (temporal) constraints and relations these tasks have and accordingly put them into a right order. (*First, I will do z, then I need to do x and afterwards I can do y.*) This will be referred to as *strategic planning*.

Of course, many future-oriented behaviors involve sequences of behaviors without being instances of strategic planning. The digger wasp effectively conducts several actions in the right order, but certainly did not think about in which order to conduct the relevant steps beforehand. Rather, her behavior is inflexible and unable to take changes in the environment into account and, thus, represents a fixed action pattern. To count as an instance of strategic planning, therefore, an individual needs to be able to mentally identify and organize the involved actions – that is, bring them into a right temporal order – rather than just execute them in a fixed order.

Strategic planning will be distinguished from *anticipatory planning*, referring to the act of *planning for* or *ahead*. Crucially, anticipatory planning, in my usage of the term, does not necessarily involve several steps, in contrast to strategic planning, as defined above. To plan ahead, it may be enough to take one single course of action, therefore, in contrast to Szpunar et al. (2014) definition.

Anticipatory planning involves the anticipation of a future *mental* state.[81] The mere anticipation of a state in the world such as an empty fridge is not sufficient for anticipatory planning, because whether or not an individual is motivated to act is dependent on whether the future state will be appreciated or not. A recourse to the familiar examples of John and Jane from chapters 4.2 and 4.3 might help to clarify this claim.

An anticipated future state can be an external state in the world – say, the anticipation of an empty fridge on a holiday, when the supermarket will be closed –, or a mental state of oneself or another person – one's own or one's

81 Note that this understanding of anticipatory planning is slightly different from Cochet and Byrne's, who define individual anticipatory planning as "[m]entally reviewing an intended action sequence to anticipate a future need" (Cochet and Byrne 2015, 239, table 1).

friend's tiredness in the afternoon, say. To become an instance of planning, however, this anticipation of a future state must be additionally 'evaluated' in the sense that the subject needs to take a motivational relation to this future state. Is John going on vacation? Then the empty fridge might be welcome. Is Jane sick and desperately needs some sleep? Then her sleepiness (maybe due to medication) may be longed for. If, however, John and Jane know that neither the empty fridge nor the sleepiness will be desired the moment they occur, this provides them with reasons to take action, that is, to plan. Thus, anticipatory planning *necessarily* involves the anticipation of a future mental state.

Of course, it is also possible to plan for a future state that does not qualitatively differ from one's present one. Take again the example of the empty fridge: maybe John's fridge has been empty for the last couple of weeks. Its future emptiness does not differ from its present status. Something else must change that gives John a reason to act now. Assume that he is on a diet that will come to an end by tomorrow. John knows that his evaluation of the fridge's status will change. Whereas he currently approves of the empty fridge, because it makes it easier for him to stick to his diet, he anticipates that he will not do so tomorrow. Thus, it is not the external state in the world that qualitatively changes, but John's mental state.

To sum up, strategic planning refers to the process of mentally representing several steps or sub-goals, at least some of which need to be brought into a temporal order, necessary to achieve a goal state, located in the non-immediate future. This instance of planning emphasizes the cognitive process of determining the correct or most efficient strategy. Daily life plans often have multiple possibilities for realization, whereas artificial planning tasks often have only one correct (or most efficient) solution.

Anticipatory planning puts more emphasis on the determination of the individual's future motivational relation to a state of the world. Thus, it involves the mental representation of a future state and, importantly, one's motivational relation to this state, such as a future need. Anticipatory planning does not necessarily involve more than one step for the successful achievement of the goal. To plan ahead, it may be sufficient to take a single precautionary action to prepare for a future contingency.

Importantly, for folk psychology it does not matter whether the current and the future motivational state of the person are qualitatively similar or dissimilar. In making plans (e.g., planning your PhD), the future goal is often currently desired by the individual. Similarly, one can plan for one's anticipated hunger tomorrow, while being hungry right now as well. We would not deny a person her planning abilities, if we found out that she was hungry when she went to the grocery store. As has been argued above, this indifference with regard to the planner's current motivational state, however, brings about certain problems for empirical verification, because the future motivational state must be clearly distinguishable from any needs or desires the individual currently has to clearly differentiate whether the individual argues in accordance with her future or merely her present one.

4.4.3 *Three Temporal Perspectives*

It is generally undisputed that associatively learned behavior or fixed action patterns, such as the maternal feeding behavior of the digger wasp, described in chapter 4.1, do not fall under the concept of planning. With regard to Santino, however, there is considerable disagreement as to whether his stone-caching behavior should suffice as evidence for his ability to plan or not. Planning presupposes the ability to anticipate the future. Up to now, it is unclear, what this exactly entails. Is the ability to represent events that have not happened yet and act accordingly sufficient? Or does a planning individual additionally need to represent these future states *as future states*?

To be able to base one's decision on this matter on reasonable ground, one central question, which has continued to emerge during the last few chapters, still waits to be answered, that is, whether planning presupposes a grasp of time and, if so, which kind of. The future can be represented rather differently, as has become evident in chapters 4.2 and 4.3. Furthermore, it is commonly acknowledged that between two and six years of age, children undergo massive changes concerning their ability to think about time. Thus, it may theoretically be the case that animals, just like small children, have great difficulties in understanding that there are often two or more ways how the future can unfold, or that they might know in which temporal order two events are going to happen, but still not understand what the temporal order tells them with regard to the future whereabouts of the involved objects, and so on.

I agree with Hoerl and McCormack (2011) that mere anticipation of a future state is insufficient to ascribe planning to an individual and that a planning animal must have *some sense of the future*. I disagree with these authors concerning the requirements such a sense of the future must fulfill, however. I will argue that anticipatory planning does not necessarily presuppose a mature human-like understanding of the future in a linear temporal framework. Instead, a representation of the future as potentially alterable will be argued to be sufficient. Strategic planning, in contrast, seems to be much more cognitively demanding and may, in fact, only be accomplished via a linear temporal framework. Teresa McCormack (2014) distinguishes three[82] temporal

82 McCormack leaves it open whether there may in fact be more than these three (McCormack 2014, 83).

perspectives children typically acquire one after the other, which will be briefly described in the following paragraphs.

4.4.3.1 Script-based Knowledge

Children as young as two seem to be sensitive to the difference between past and future events, because they can correctly answer what just happened or what is going to be the case. Their temporal perspective is dually limited, however. For one thing, these children are only able to recall or predict events that are part of what has been called *scripts*. Scripts allow a child to know what is to come in repeated or familiar settings, during bedtime routine, for example, or dinner at a restaurant. Jane, for instance, knows that the main course will be served after the appetizer. In order to generate a script, an individual must have experienced or observed several more or less similar sequences of events. Consequently, script-knowledge is extremely limited in scope, because it cannot inform the individual about future events in unfamiliar settings.

Moreover, event sequences in a script are represented in cycles, rather than linearly. This means that an individual with script knowledge can make judgments about temporal order within *one* script, but she cannot separate two occurrences of a similar event, for instance, yesterday's going-to-bed and tomorrow's going-to-bed. Also, two events of two different scripts cannot be brought into one temporal order, which means that a child may not be able to know which one of two events will precede the other. This being said, there may exist meta-scripts, however, which could, in principle, integrate events from two or more separate scripts into a third. Even in such meta-scripts, however, event sequences would still be represented in cycles.

In spite of these limitations, however, two-year-olds are often able to answer which event will unfold next, that is, they are seemingly able to anticipate a future state. It appears, therefore, as if they must be able to orient themselves within a script and grasp the temporal relations between its events. John Campbell indeed was certain that even very young children already possess a basic understanding of the past and the future (Campbell 2006).

There are reasons to doubt that this perspectival way of thinking should be considered to involve a *temporal* perspective at all, Hoerl and McCormack (2011) argue that aspectual notions of events (such as, 'completed', 'ongoing', 'yet-to-come') must be differentiated from temporal relations (for instance, 'before', 'after'). The authors hypothesize that before acquiring a knowledge of temporal relations, children possess an understanding of aspectual notions. This grasp of aspectual notions enables even a very young child to anticipate certain events and even to prepare accordingly (bring a book to bed, for instance, in expectation of a bedtime-story). In an important sense, however, such a child is still stuck in the present in that she cannot mentally transcend her present perspective:

> [I]f a young child is thinking of events only in terms of aspectual notions, there is a sense in which her focus is restricted entirely to the present itself. That is, she does not think of past and future events in their own right, as located at other points in time, but only about their current status as completed, ongoing, or yet-to-start. To transcend the perspective of the present, she would also have to be capable of thinking, say, of events she now thinks of as completed as having once been ongoing, or of events she thinks of as ongoing as completed in the future. (Hoerl and McCormack 2011, 447f)

Restricted to the present perspective, the ability to anticipate future events via script knowledge involves temporal updating, which has been laid out in detail in chapter 4.2.3. As new information is added, the individual constantly updates her model of the world. Thus:

> [L]earning a script, at least at an early age, is not so much a matter of coming to think of a sequence of events *as* unfolding in a certain order, but rather a matter of coming to think of those events *in the right order* [...]. On this view, there is no overall representation of the sequence in which the child then orientates herself with the help of token-reflexive notions. Rather, recounting a script, at least initially, actually involves representing those different events in sequence. (ibid. 447)

If Hoerl and McCormack are correct, then script knowledge, at least initially, cannot provide any information about time at all and should possibly not be considered as a *temporal* perspective, consequently. I take their claim to be plausible, even more so, because a minimal requirement for a grasp of time seems to be that the past and the future categorically differ from one another. If the sole possession of scripts, however, does not allow an individual to distinguish between two occurrences of going-to-bed, where one lies in the past and one is yet to come, then it seems strange to ascribe that individual with an understanding of time.

4.4.3.2 Alterable and Unalterable Events

At about three years of age, children acquire a knowledge of the distinction between the past and the future that seems to go beyond scripts, in that they can now differentiate between unalterable (that is, past) and potentially alterable (that is, future) events. In detail, they learn to refer to those events, even though they are not connected through scripts to the present and they begin to provide descriptions of particular scenarios (e.g., a specific visit at a restaurant) rather than general scripts of event sequences (McCormack 2014, 85).

This is not to conclude, however, that three-year-olds have a mature systematic understanding of time. Rather, it is plausible to assume that three-year-olds can distinguish between past and future events in terms of their (un) alterability, without also being able to think separately of their unique points in time. Thus, children at this developmental stage can differentiate one future event x from another future event y, but may still be unable to reason about their relative locations within a linear temporal framework. Similarly to script-based thinking about events with regard to their aspectual notions, this immature way of thinking "involves thinking of events as having a certain status, rather than thinking of points in time per se" (ibid. 86).

4.4.3.3 Event-independent Thinking in a Linear Temporal Framework

McCormack suggests that until now, children have only been able to engage in *event-dependent* thinking, that is, they can represent points in time only by reference to their specific corresponding events. *Event-independent* thinking, in contrast, allows an individual to think about time itself and represent temporal locations as empty slots, where alternative events could (have) take(n) place (McCormack 2014, 86f). In this way, event-independent thinking enables hypothetical and counterfactual thinking. John, for example, knows that in the afternoon, he could go grocery-shopping, read a book or do something completely else. Similarly, John knows that yesterday he could have gone to the gym, even though he did in fact go to the library.

What is more, John knows not only that the event of him going to the library is not alterable anymore (because it is in the past), but also that it had been alterable at some point in time (because it still was in the future). Because of this, it is plausible to assume that in order to engage in event-independent thinking, an individual must necessarily possess a temporal framework representing time on a linear timescale, stretching from the past to the present and into the future. Such a linear conception of time allows an individual to understand exactly that as he moves through time, events that only remain future possibilities now will have transformed into unalterable past events tomorrow.

I have argued that event-independent thinking plausibly requires thinking in a linear temporal framework. It might seem that the opposite is not automatically true. In principle, it seems that an individual could very well be able to locate events on a linear timescale without also being able to engage in event-independent thinking, that is, understand that at a specific location in time, y could happen instead of x, for instance.

If, however, a linear temporal framework is understood as to necessarily involve the ability to transcend the present perspective in that it allows the individual to understand that what now is in the past has once been in the future, then a linear temporal framework and the ability of event-independent thinking are inseparably connected to one another. The moment a child not only grasps the difference between past and future in terms of alterability, but also understands that events, which are *now* unalterable have *once* been alterable, she possesses the ability to engage in event-independent thinking and to think about time in a dynamic, linear temporal framework. With regard to McCormack's analysis of the development of temporal thought, the child has acquired the third and last stage.

Children are able to engage in real hypothetical and counterfactual thinking only at about five years of age, so it seems reasonable to suggest that they acquire a mature temporal perspective at around the same age. Importantly, the ability to answer standard – that is, closed – hypothetical and counterfactual questions seems to develop earlier than the ability to represent branching futures (for details, see chapter 4.2.4). Even though young children, for instance, can imagine a counterfactual world, they do not yet fully seem to understand that this imagined scenario represents an *alternative* scenario to the one that *actually* did happen. Similarly, they can answer standard hypothetical questions without also representing the temporal-causal relationship between current reality and their mental representation. Therefore, the ability to imagine scenarios that differ from current reality is not *per se* sufficient to show that an individual has gained the capacity to think about time event-independently.

Because it enables hypothetical thinking, event-independent thinking is particularly important for simulation. With Szpunar, it has been suggested that an individual, who can simulate the future, that is, mentally represent a sequence of future events, must also be able to distinguish possible from impossible events. As has already been laid out in chapter 4.3, this can be interpreted to mean, first, that only realistic or statistically probable scenarios are being taken into consideration – which only required the capability of mental space travel – but it could also be interpreted as to include temporal constraints. Thus, John can hypothetically think about multiple different but possible ways how tonight's dinner will unfold, because he knows, for example, that if he chooses to serve the rosé as aperitif, he will no longer be able to use it as accompanying wine for dessert later; but that if he wants to serve a cocktail as aperitif instead, he will need much more time for preparation and will not be ready with dinner when his guests arrive. In order to engage in this kind of simulation, John must be able to think about the future in an event-independent kind of way and possess the ability for mental time travel.

4.4.3.4 The Temporal Perspective Required for Planning

Having sketched the developmental changes in temporal thought, this chapter aims to finally answer the question how much of an understanding of time planning actually requires. I take it to be consensual that planning presupposes an understanding of temporal rather than aspectual notions. If this is acknowledged to be true, then planning goes beyond the quasi-temporal perspective acquired through scripts. Two stages of temporal perspective remain to be assessed, therefore.

As the example of John has shown, hypothetical thinking is essential for strategic planning, where the agent mentally simulates alternative event sequences and their consequences in advance of taking action. It seems reasonable to assume, therefore, that event-independent thinking is a necessary requirement for the ability to engage in strategic planning and can only be performed by those individuals who have acquired the third and last stage of temporal perspectives.

What about anticipatory planning? In order to be ascribed with planning abilities at all, Hoerl and McCormack argue that an individual must necessarily be able to engage in event-independent thinking, empirically accessible through hypothetical and counterfactual thinking. Thus, they consider the final of the three temporal perspectives as a general requirement for planning. Put differently, even though Hoerl and McCormack surely agreed that anticipatory planning behavior *can* be a manifestation of the underlying capacity to plan, it is not necessarily so. Crucial for the ascription of planning abilities is, according to the authors, the individual's capability of event-independent thinking and thus, as has been argued above, an underlying linear grasp of time as stretching from the past to the present and into the future.

It might be questioned whether anticipatory planning actually requires such a mature understanding of time as well. In fact, I will suggest that the event-dependent temporal perspective, which amounts to the second stage in McCormack's developmental analysis, should be considered as sufficient to ascribe an individual with the ability of anticipatory planning.

Additionally to a notion of the alterability of the future, an individual with anticipatory planning abilities must know that she herself can affect future outcomes. This means, that the future is not only potentially alterable but is in fact alterable *by the agent herself.*

One way to do this is to episodically simulate an *undesired* scenario and define an alternative desired goal state. This would in fact presuppose event-independent thinking, because the individual needs to understand that her mental representation of the future scenario is just one alternative of how the future could theoretically unfold. Thus, the episodic simulation in this example involves mental time travel, because it involves a grasp of temporal-causal relationships. In simulating the embarrassing situation of being unable to serve any dessert but also knowing that the future is alterable, John forms the intention to go to the supermarket. Without the knowledge that the future is alterable by himself, he would still be able to anticipate the undesired future scenario but could merely 'brace himself' for the uncomfortable situation to arrive, unable to understand that the anticipated outcome can still be changed.

This does not mean, however, that without mental time travel capacities an individual is doomed to passively wait for whatever is going to happen. In cases, in which the individual simulates a *desired* scenario, the representation can function as goal state and motivate the individual to act in an adequate way to achieve the mentally represented goal state. In fact, Redshaw hypothesizes that environmental cues can 'trigger', via mental space travel, a certain adaptive future-oriented behavior.

Consider again Redshaw's (2014) example of the hungry chimpanzee, coming across some stones. Redshaw proposes that this sight causes in the animal's mind a representation of himself opening nuts, so that the chimpanzee is motivated to take these tools with him. Importantly, Redshaw argues, the chimpanzee does not have to be ascribed with a metarepresentative understanding of the specific temporal context. I want to argue that even if this might be true, it does not rule out the possibility that the chimpanzee at least understands that taking the stones with him will result in him eating nuts, while leaving the stones behind will result in him not eating nuts. Put differently: whether an individual, simulating a scenario by means of mental space travel, understands that a certain action or event will result in the desired outcome, or whether the mental representation merely initiates or 'triggers' a certain adaptive behavior is written on another page.

To further clarify this, consider again children's development of hypothetical and counterfactual thinking abilities. Before acquiring an understanding of branching futures, children seem to evolve a grasp of the alterability of the future, because they easily imagine counterfactual and hypothetical scenarios, even though they seem to be unable to fully understand the nature of the relationship between these representations and current reality. It seems plausible that the ability of event-independent thinking evolves gradually, so that an individual might already possess some basic understanding of hypotheticals and counterfactuals, without simultaneously being able to represent branching futures, and, thus, in the absence of a full-fletched linear temporal framework.

In cases, in which mental space travel results in a causal (but not fully developed temporal-causal) understanding of the relationship between two events, I take it to be plausible to assume that the future is being represented as alterable. Anticipatory planning does not necessarily presuppose event-independent thinking, therefore. If a certain adaptive behavior is merely 'triggered', however, I am more reluctant to call these instances planning, contra Redshaw (2014), because, if the meaning of 'triggering' is taken seriously, there should be no understanding involved of how exactly one's present actions will affect the future outcome. To sum up, if the chimpanzee from Redshaw's example indeed understands that taking the stones with him will enable him to eat nuts, whereas, if he does not take the stones with him, he will not be able to do so, this seems sufficient to assume that he has a grasp of the future in terms of its alterability.

Of course, to determine whether a particular behavioral outcome should be considered as the result of mental space travel merely triggering a certain adaptive behavior or whether it should rather be acknowledged as an instance of anticipatory planning, should be difficult. However, as will be seen in chapter 5.1, certain methodological requirements have been proposed for empirically testing anticipatory planning abilities in animals, with the aim to exclude alternative non-planning explanations, such as the one above. If, for instance, Redshaw's stone-gathering chimpanzee could be shown to show a similar seemingly planful behavior in other situations and, importantly, other domains as well, this would strengthen the anticipatory-planning explanation, because it would be quite implausible to assume that for each and every kind of mental simulation, a certain adaptive, goal-oriented behavior could be triggered without the animal actually planning ahead to the future situation itself.

To summarize, I have separated two types of planning, anticipatory planning and strategic planning. Strategic planning arguably presupposes event-independent thinking in a linear temporal framework, represented by the third stage in McCormack's analysis of temporal thought. Event-independent thinking allows a representation of time as empty slots, which can be occupied by alternative events. It enables hypothetical thinking and counterfactual thinking as well and is, thus, a requirement for mental time travel and temporal-causal reasoning.

With regard to anticipatory planning, it has been suggested that an understanding of the future as potentially alterable by one's own actions might be sufficient. Thus, anticipatory planning in many cases only requires an event-dependent temporal perspective that involves a grasp of the categorical distinction between past and future events in terms of their (un)alterability, that is, the second of McCormack's three temporal perspectives.

Anticipatory planning can occur in the absence of episodic simulation, of course, in that not every instance requires mental space travel or mental time travel. It should be noted, however, that script-based knowledge alone is insufficient for anticipatory planning, because it does not fulfill the minimal requirement of involving temporal rather than aspectual notions. Whether an individual can plan for her future in the absence of an episodic memory system, thus, is to ask, among other things, whether the semantic memory system alone can provide an individual with the categorical distinction between the past and the future.[83] Up to now, I take this to be an unresolved topic, even though more research might very well provide answers in the future.

4.4.4 *A Closer Look on the Components of Planning*

The following chapter takes a closer look on the planning process itself. This enables me to formulate the challenges any empirical investigation of animal planning capabilities faces and to point out more differences between anticipatory and strategic planning. There are different approaches for an analysis of the components of planning. One way would be to differentiate cognitive planning from planning behavior. It might be a mistake, however, to treat them separately, because cognitive planning and the related behavior often are not distinguishable. Plans are seldom formulated first and then executed; rather, cognition and action often go hand in hand and intermingle. For this reason, planning will be regarded as one (cognitive-behavioral) process.

According to Ellin Kofsky Scholnick and Sarah Friedman, "[p]lans and planning [can be] described as a set of complex conceptual abilities that reflect knowledge, the ability to represent it flexibly and abstractly, the ability to recognize or set goals, strategic skills, and skills in monitoring, evaluating, and repairing strategies" (Scholnick and Friedman 1987, 29). The authors distinguish between six main elements that are involved in a planning process, which will be examined more closely. In the following paragraphs, it will become evident along the process that strategic and anticipatory planning mainly differ from their emphases of different components of planning.

83 As mentioned previously, hippocampus-lesioned patients are severely limited in their ability to engage in episodic simulation, because of their deficits in mental space travel. Still, however, they seem to have retained their temporal-causal reasoning abilities and their experience of time in a linear framework, which should still make them able to plan ahead, at least in principle. Interestingly, however, this is often not the case. Szpunar (2010) suggests that this might be due to their inability to mentally put individual components together into one unitary novel scenario, rather than an inability to think about future *per se*, however. Thus, hippocampus-lesioned patients may very well be able to anticipate future events and have a preserved grasp of the distinction between past and future events, but still not be able to plan ahead.

4.4.4.1 The Representation of the Problem Space

A prerequisite of planning is an adequate mental representation of the environment. This representation is not a replica of the perceived world, but depends, among other things, on the subject's understanding of the spatial, temporal and causal relations of its involved objects. The specific appearance and quality of these abilities thus greatly influences the individual's planning abilities (see also chapter 4.2 and 4.3). Especially when the goal is not pre-determined by the experimenter (as is the case in many anticipatory planning tasks, where the identification of a goal state is already part of the planning process itself), individual differences in representational skills can have large impacts.

To illustrate the role of problem representation, take the example of an anticipated sleepiness in the afternoon and imagine further that you are at a conference. Assume you have just finished your talk. Even though you are wide awake right now, rushing with adrenaline, you anticipate that you will get tired pretty soon, because you stayed up all night, preparing your presentation. Your hotel room is far away and not an option.

Now, there may be two different ways to take action for you, depending on the need or desire you choose to act on. You might plan ahead for your sleepiness by taking a cushion with you wherever you go. When the fatigue will become overwhelming, you can coil up on the carpet and take a nap during the conference. Put differently, you take precautions to meet a future need.

Alternatively, you could anticipate that you will not want to be tired in the afternoon and, based on this thought, might try to change your anticipated sleepiness. A reasonable thing for you to do, then, would be to position yourself near an open window and the coffee machine (even though you do not feel a bit tired right now). These two strategies represent very different ways to handle the anticipation of a future state that will be undesired. In the first example, it is the situation that is considered to be undesired (wanting to sleep in an uncomfortable room), whereas in the second example it is the wish to sleep itself. The latter is closely connected to autonomy, defined as second-order capacity to critically reflect on one's first-order desires. Depending on her cognitive capabilities, two individuals might have two very different representations of the problem space and might pursue very different planning strategies, therefore.

4.4.4.2 Goal Definition

In order to plan, the subject must choose a goal. In empirical investigations of strategic planning, the objective of the task is often well-defined, because it is clearly stated by the experimenter. The planning part comes in later, with the mental identification and organization of the necessary steps that will lead to the achievement of the goal and the execution of the successful strategy. There may be only one or several possible solution. Sometimes, the task requires identifying the most efficient among several possible solutions, as is typically the case in Tower of Hanoi[84] and similar tasks.

In anticipatory planning, the identification of the goal is already part of the planning process: to plan ahead, the subject must show that she can anticipate her future motivational mental state (*I will be hungry*) and accordingly define a goal (*food*). In actively pursuing this goal state (by going grocery-shopping, for instance), the individual is commonly ascribed with an underlying ability to plan for a future state, if further requirements are met as well (see also chapters 4.2, 4.3 and 5.1).

Especially in situations where the goal is ill-defined (because, for instance, there are multiple goals involved or the goal itself is vague), the identification of a goal can be quite complex, with the goal(s) being monitored, evaluated and eventually modified during the execution (Scholnick and Friedman 1987, 10–13). Moreover, the problem space in real-life settings is not static, so that an advanced planner should be able to flexibly adjust her goals and strategy along the way.

4.4.4.3 Deciding to Plan

After the individual has set a goal, but before formulating a plan on how to achieve this goal, she must decide to plan in refraining from immediate action. Scholnick and Friedman (1987) call this willingness to engage in planning *planfulness*[85]. Planfulness obviously requires self-control but is influenced by other aspects as well. Self-control and some other important aspects will be shortly sketched in the following paragraphs.

Especially in situations for which the ability of anticipatory planning is relevant, the individual must often exhibit a whole lot of self-control, even more so in delayed-reward experiments, where the subject is being tempted

84 The game of the Tower of Hanoi consists of three poles and a number of disks of different sizes, that are all stacked in ascending order of size on one pole. The player is only allowed to move one disk at a time and is only allowed to put a disk on top of a bigger but not a smaller one. The objective of the game is to rebuild the stack of disks as it appears in the beginning of the game on another pole in the fewest moves possible. In the related Tower of London test, participants are typically required to move three colored disks of equal size on three poles and to rebuild a specific target position in the fewest possible moves (Shallice 1982). I will come back to these tasks in chapter 5.2.

85 Note that this usage of the term is slightly different from how it has been used previously. Above, I repeatedly used the term planful behavior when referring to the kind of behavior that could be read as involving planning on behalf of the individual in contrast to a behavior that does not involve planning.

to take an immediate, smaller treat, which would deny a larger, future reward. Self-control is not only sensible to situational aspects, however, but also to species characteristics as well as personal traits. Certain species and individuals, respectively, thus, will be expected to perform better than others in self-control experiments.

For another thing, a planning individual must understand that she is a causal effect in the world. In this context, this is supposed to mean that additionally to the more physical understanding of causality, represented, for instance, in her causal powers reasoning, the individual must also psychologically believe that she is the author of her own future, at least to a certain extent. This aspect may be particularly important for comparative cognition, because a captive animal, even if she has, in principle, a grasp of how her actions influence future outcomes, might get the impression that what happens to her often lies outside her own responsibility.

Take the example of a zoo chimpanzee hurting her ankle. Even though no other individual (but a camera) observed the accident, a man in a white coat appears shortly afterwards and immediately sets out to investigate the animal's foot. From this and lots of other similar experiences, the chimpanzee might come to the conclusion, thus, that humans do not depend on observing events for themselves, but are omniscient. Similarly, captive scrub jays, discovering that their caches have been emptied, or Santino, whose hidden projectiles were regularly removed, may have a disturbed understanding of causality and their own causal influence on their environment, which might sometimes negatively affect results in planning experiments. The risk of a possible influence of human interaction on captive animals' planning abilities can only be rationally estimated by those working with the animals themselves, of course.

Planfulness also requires the individual to know under which circumstances planning is sensible and when it is not. Routine behavior in humans (e.g., driving to work) very often does not require complex plans. If a certain action sequence is repeated over and over again, it becomes at least possible, therefore, that the individual learns to execute the relevant steps without engaging in complex planning. This is why planning experiments are typically careful not to conduct too many trials on similar tasks and to test for ecological or behavioral predispositions, that is, domain-specific evolved adaptations that do not require planning, but which can facilitate rather complex behaviors nonetheless (see also Kabadayi and Osvath 2017).

At the other extreme, it may be equally unwise to put much brain energy into the formulation of a full-fletched strategy, if doing so is cognitively very demanding (because there are many alternatives, for example), the planning process will undergo many changes (because it involves uncontrollable factors) and/or the additional effort of developing a strategy simply is not worth the pain, because mistakes will not have extreme negative consequences (Scholnick and Friedman 1987, 16). Thus, it can be wiser to decide not to plan.

Clearly, this presents several challenges to animal researchers. First, the problem that the animal is required to solve, must be so demanding as to inhibit the impulse to take immediate action and rather motivate the animal's willingness to engage in planning. Furthermore, the consequences in case of failure need to be so high as to motivate the animal to engage in planning, on the one hand, while also making sure, on the other hand, that the task is not too complex as to preclude the attempt to mentally come up with a solution first. Also, as mentioned above, the behavior must not become routine for the animal along the way or test a kind of behavior, for which the animal might already have a predisposition (e.g., a fixed action pattern).

Taken together, these facts require finding a delicate balance on behalf of the experimenter between routinely and predispositional behavior, on the one hand, and excessive demands, on the other hand. To meet these demands, certain methodological requirements have been proposed that should be fulfilled by planning experiments. The extent of planfulness seems to depend on several aspects. Personal beliefs, character traits, cognitive skills as well as cross-species differences but also situational settings all seem to determine an individual's ability to refrain from immediate action in a concrete situation (Scholnick and Friedman 1987, 13–19).

4.4.4.4 Formulating a Plan

Once the individual has decided to plan rather than jump into action immediately, she needs to identify and organize the necessary steps in order to achieve the goal. In case of anticipatory planning, this may be relatively easy, because the course of action often includes only one single step. This is different in case of strategic planning, however, where several individual actions must be carefully mentally arranged.

Scholnick and Friedman differentiate anticipatory plans from opportunistic planning. To avoid misunderstandings with what I have called anticipatory planning, I will distinguish between *opportunistic* and *non-opportunistic planning*, however. Planning is often thought of as a neat 'first-think-then-act' process, involving the mental predetermination of a complete action-sequence that will lead to the achievement of the goal. Planning, on this approach, is top-down: first, the goal and the subgoals are defined, then all the single steps are specified and mentally brought into the right order, and only then the subject executes the necessary actions. Neat top-down plans work best when the goal is well-defined, the problem space is familiar and the problem itself is relatively simple (ibid. 19f).

However, because animals as well as humans are equipped with only limited working memory capacities[86], it might not always be feasible for them to engage in this kind of non-opportunistic planning and might, even worse, result in the abandoning of those aims that are rather ill-defined, stretch into the very distant future, have multiple alternatives for realization and so on. If the problem space furthermore includes unpredictable changes in the environment, non-opportunistic planning may not be an ideal option or too time- and energy-consuming to be of practical use to the individual (see also Hayes-Roth and Hayes-Roth 1979).

For these reasons, a plan alternatively not be formulated in advance but in an opportunistic way. In contrast to the 'first-think-then-act' motto of top-down planning, opportunistic planning can be described as "plan[ning] in action" (Scholnick and Friedman 1987, 21). Rather than mentally trying out the complete course of action prior to execution, the opportunistic planner anticipates a required course of action while also simultaneously executing certain actions and also monitoring them. The boundaries between the mental act of formulation and the behavioral act of implementation thus blur. It is crucial to keep in mind, however, that even in opportunistic planning, planning still emphasizes the mental act of trying out certain acts and consequences *beforehand*. Similarly, Hayes-Roth and Hayes-Roth describe planning as "the predetermination of a course of action aimed at achieving some goal" (Hayes-Roth and Hayes-Roth 1979, 275f). Thus, opportunistic planning is not to be confused with trial-and-error learning or a reversal of the non-opportunistic planning motto into a 'first-act-then-think' bottom-up method. In the following paragraphs, I will attempt to further clarify opportunistic planning, as it will be understood here and hereafter.

As has been mentioned above, opportunistic planning can be especially valuable in rather unpredictable or very complex settings. But especially when a first step into the wrong direction can potentially lead to the failure of the whole project, the individual is well-advised to consider all possible, or at least all presently relevant, possible courses of action against one another and only then execute her first move. A proficient planner, thus, should be able to recognize whether a task necessarily requires a top-down planning approach or whether not each and every possible course of action must be mentally tried out prior to taking action.

---

86 Note that a top-down plan may be extremely efficient for artificially construed systems with a very large working memory system.

It will be assumed that strategic planning may, in theory[87], occur in both ways, that is, in a strict and neat top-down manner or in the form of opportunistic planning. This means that the individual is not necessarily required to mentally try out the *complete* strategy beforehand or that she must be able to represent *every* possible alternative course of action that there is. Because event-independent thinking has been defined as a requirement for strategic planning, however, the individual must be able to mentally fill out a time slot with alternative actions or events, that is, she must be able to engage in hypothetical thinking.

4.4.4.5 Executing and Monitoring a Plan

Even during the execution of a plan, external or internal changes, for instance, in the environment or the motivation of the individual, are constantly monitored. Monitoring makes sure that the subject comes closer to the goal state and that the plan is repaired, modified or made more efficient, if necessary. A certain step, initially being part of the strategy, may have become obsolete, for example, or another step might alternatively need to be included. Monitoring becomes extremely important in opportunistic planning, where no ready-made plan is available, but where after each step, the situation is re-evaluated.

4.4.4.6 Learning from Plans

Scholnick and Friedman (1987) argue that there are at least two possible ways how the subject can learn from planning: the applied strategy may become routine after some time (consider again the car drive to work, which does not require complex future-thinking abilities after several repetitions) and thus making the energy-consuming process of planning superfluous. It is also plausible to assume that proficient planning abilities might foster other competences in other cognitive domains as well.

Above, I have argued for a conceptual separation between two anticipatory and strategic planning. Planning always involves the anticipation of a future state, but depending on whether more emphasis is laid on the definition of the goal state itself, as in anticipatory planning, or the cognitive process of bringing several individual steps into a correct temporal order, as in strategic planning, planning tasks can considerably differ from one another. Strategic planning,

---

87 It is another question which one of the two models might actually better capture the relevant cognitive and behavioral processes in humans and animals. However, as even proponents of the opportunistic-planning model such as the Hayes-Roths emphasize the role of anticipation and a predetermination of relevant steps for planning, both models are relevantly similar with regard to my research topic.

with its requirement of the ability to engage in temporal-causal reasoning (event-independent thinking), is probably cognitively more challenging than anticipatory planning. An individual capable of anticipatory planning does not require the ability to engage in event-independent thinking. In contrast, the second of McCormack's three temporal perspectives should be enough, that is, the individual must at least grasp the difference between future and past events in terms of their (in-)alterability.

As a result of the different requirements of anticipatory and strategic planning, Scholnick and Friedman's (1987) six major components of the planning process acquire varying significance for an empirical investigation of planning. Below, the empirical evidence for anticipatory and strategic planning abilities in animals will be finally examined and afterwards discussed with regard to their impact on McMahan's prudential unity relations.

CHAPTER 5

Planning Animals?

5.1 Behavioral Investigations of Anticipatory Planning in Animals

5.1.1 *The Spoon Test Paradigm*

For a long time, empirical research has largely focused on anticipatory planning abilities, which can be attributed to a great part to the postulation of what would be come to known as the *Bischof-Köhler hypothesis*. The Bischof-Köhler hypothesis was introduced by Thomas Suddendorf and Michael Corballis in their seminal paper on the evolution of foresight in 1997 (see also Suddendorf and Corballis 2007) and has sparked a lively discussion on future thinking abilities in animals. Named in reminiscence to the researchers Wolfgang Köhler, Doris Bischof-Köhler and Norbert Bischof, the Bischof-Köhler hypothesis states "that animals other than humans cannot anticipate future needs or drive states, and are therefore bound to a present that is defined by their current motivational state" (Suddendorf und Corballis 1997, 150). Animals, on this account, are thus unable to plan ahead and incapable of what has been called anticipatory planning.

Since Suddendorf and Corballis' postulation more than twenty years ago, the falsification of the Bischof-Köhler hypothesis has been described as the "holy grail" (Raby and Clayton 2009, 924) of animal prospective cognition research. Whereas Corballis (2013) recently distanced himself from his and his colleague's original position, Suddendorf has repeatedly reinforced the hypothesis that only humans can mentally transcend the present. Together with Endel Tulving, he now represents the main counterpart in the debate around planning abilities (and episodic cognition) in animals. On the same time, however, Suddendorf and Tulving have encouraged empirical researchers to empirically falsify their hypotheses and suggested experimental set-ups to do so. The progress, that has been made in the area of comparative prospective cognition research during the last decades, therefore, can be attributed to a not-negligible part to the fruitful discussion between proponents and opponents of the Bischof-Köhler hypothesis.

Under the influence of the Bischof-Köhler hypothesis, Endel Tulving (2005) proposed a test, which he termed the *spoon test*, as a means to specify the competence criteria of the allegedly uniquely human ability to plan for future needs. The spoon test is based on an Estonian children's story, in which a girl dreams about being at a party where chocolate pudding is served. In lack of a

 | DOI:10.30965/9783957437365_006

spoon, she is unable to eat it. The next evening, thus, the girl takes a spoon with her to bed with the purpose to be better prepared next time. Analogously to the planning behavior of the girl, Tulving argued that an animal who can plan for her future needs, should bring a tool in conscious anticipation of a future event.

By itself, the spoon test cannot measure whether the action is motivated by the girl's current desire for chocolate pudding or by her knowledge that she will develop a desire for pudding later. This difference is crucial, however. In the first case, the girl does not necessarily anticipate a future state at all, similar to the situation, when John goes grocery-shopping while being hungry. If planning behavior is to be separated from other intentional or goal-directed behavior, however, then all those cases where the present and the future motivational states do not qualitatively differ from one another should be discarded (see chapter 4 for details). Applied to the Estonian girl, this would require to change the story insofar that the girl should still take put a spoon under her pillow when going to bed, even though she currently had no craving for chocolate pudding.

Besides the requirement that the behavior must not satisfy a present need or otherwise be explainable by mechanisms[88] other than the conscious anticipation of a future motivational state, spoon tests must fulfill two more requirements, according to Tulving. To rule out associative learning or other subconscious retrieval mechanisms, the experimental situation must exclude possible triggering of the behavior in question by environmental cues that were present when learning initially took place. This means that for testing, the individual should be re-located into another environment, that differs from the original learning set-up.[89] Finally, the behavior must satisfy a need that will predictably arise in the future to make sure that the individual's mental representation actually refers to an event that is yet to come (Tulving 2005, 44f).

Naturally, spoon tests can be varied to a great extent, depending on, for example, the specific *umwelt* of the relevant species. They may not always involve the usage of a tool, for example, but rather ask for other preparatory

88 Tulving rather vaguely mentions a behavior "governed by current physiological states" (Tulving 2005, 44), that must be excluded.

89 Suddendorf and Corballis (2010) additionally mention that the two instances of the preparatory behavior and the future event, for which the animal must plan ahead, must be sufficiently spatially and temporally separated. The reason for this is to ensure that the relevant behavior, first, actually represents a simulation of an event rather than being attributable to spatial cuing, and, second, draws on long-term memory rather than working memory. This is an important but rather uncontroversial condition, which is generally not being violated (but see Bräuer and Call 2015), so I only mention it as a side note here.

behaviors. If the three requirements above are being met, a passed spoon test is argued to count as sufficient evidence against the Bischof-Köhler hypothesis and, thus, in favor for the claim that animals are indeed able to anticipate and plan for future needs or desires.

In spite of its positive influence on the progress of the research area on animal planning abilities, the requirements of the Bischof-Köhler hypothesis remain controversial. Mathias Osvath recently criticized both the Bischof-Köhler hypothesis and the spoon test paradigm as “rather murky” (Osvath 2016, 8) and argued that the debate on whether the Bischof-Köhler hypothesis has been falsified by spoon test experiments or not, has been led rather heatedly, but not very constructively, leaving fronts to harden on both sides. I agree with this assessment. Whereas Osvath mainly criticizes the spoon test's requirements as a manifestation of what Buckner (2013) terms anthropofabulation, my concern is that the Bischof-Köhler hypothesis, together with the spoon test paradigm, cements the vague and incorrect dichotomy between those individuals, on the one hand, who are stuck in time, unable to anticipate anything other than what they currently perceive, and, on the other hand, the mental time travelers with a linear temporal framework.[90] The “murkiness” of the Bischof-Köhler hypothesis and the spoon test, I will argue,

---

90 I would like to single out one of Osvath's objections here, because it relates to the general statement I want to put forward as follows. Tulving and Suddendorf emphasize that the individual's behavior in question must not satisfy a present need, but a future need or desire. Osvath now argues that it seems strange to assume that individuals can be motivated by something they currently do not desire. Rather, future desires should be understood as “current motivations or desires for future states beyond the [...] current sensory moment” (Osvath 2016, 8). Osvath leans towards an understanding of episodic simulation in terms of mental space travel, emphasizing its function as a motivator in the sense described by, e.g., Boyer (2008) and Gilbert and Wilson (2007). By bringing the future into the present, episodic simulation can result in the preference of a larger, but temporally distant, reward over a smaller, immediately available reward.

Besides an episodic simulation via mental space travel, however, some individuals can be plausibly said to additionally be able to engage in episodic simulation via mental time travel (see chapter 4.3 for details). The function of mental time travel is not so much the ability to be motivated by what is currently not desired, but rather a result of event-independent thinking, i.e., the ability to consider a temporal point in time separately from its occupying event. Mental time travel, thus, enables an individual to consider one event from more than one temporal perspective. Such an individual can truly transcend her current temporal perspective and engage in, e.g., anticipated regret. The anticipation of a regret can motivate John to go grocery-shopping, even though he absolutely would prefer reading a book instead, because he knows that he will come to regret his poor decision. Whether John explicitly episodically pre-feels the negative emotion of regret or, alternatively, just semantically knows that this will be the case, is not decisive here. Important is, rather, that there is something that goes beyond episodic future thinking is mainly due to the vagueness of the underlying abilities that the Bischof-Köhler hypothesis and the spoon test attempt to find evidence – or rather lack of evidence – for.

First and foremost, the passing of a spoon test *per se* does not allow any inferences to the underlying mode of simulation, because it cannot detect and separate semantic and episodic forms of future thinking. This is problematic, however, because the spoon test was designed with the purpose to unequivocally demonstrate the supposedly uniquely human ability of episodic future thinking in animals (Tulving 2005). In a later version of the Bischof-Köhler hypothesis, Suddendorf and Corballis further emphasized this in arguing that only humans "can *flexibly* anticipate their own mental states of need and act now to secure them" (Suddendorf and Corballis 2008, 31; my emphasis). The expression "flexibly anticipate" is consensually taken to refer to the requirement of a simulation of the future via episodic future thinking rather than fixed action patterns, associatively learned behavior or semantic future thinking.

Moreover, it has been unclear throughout the debate exactly which kind of grasp of time animals passing the spoon test are required to have and, thus, which kind of episodic simulation should count as evidence for a falsification of the Bischof-Köhler hypothesis. Sometimes proponents and opponents alike seem to emphasize episodic simulation in terms of mental space travel, at other times they explicitly restrict episodic future thinking to mental time travel including a mature grasp of time. Depending on whether Tulving's requirements are interpreted to refer to the ability to mentally represent and prepare for scenarios other than the currently experienced or, in contrast, to the metarepresentative ability to consider the relationship between two representations, an empirical study along the lines of the spoon test can paradoxically be taken as evidence for and against the Bischof-Köhler hypothesis at the same time, therefore.

As has been argued in chapters 4.2 and 4.3, there is no simple dichotomy between an inability to anticipate the future, on the one hand, and mental time travel, on the other hand. For instance, an individual who can anticipate future states may be able to semantically, but not episodically, simulate the future. A second individual may be able to engage in mental space travel, but not mental time travel. And finally, as seems to be the case in some hippocampus lesioned amnesia patients, an individual may even have a linear temporal framework and be able to form metarepresentations, but be extremely limited

via mental space travel, namely the ability to represent the temporal-causal nature of the relationship between a mental representation of the future and current reality.

in her mental space travel abilities, that is, have great difficulties to mentally put mnemonic components together in order to simulate a novel scenario. Such an individual is severely limited in her ability to form mental representations, which is a requirement for both mental space and mental time travel, but does not seem to be impaired in her grasp of time as passing from the past to the present and the future.

To conclude, the spoon test *per se* is not suitable to separate semantic and episodic forms of future thinking in my view. The spoon test paradigm *can* provide evidence for anticipatory planning, however, if it additionally makes sure to exclude a temporal perspective based on entirely aspectual rather than temporal notions, represented by McCormack's first, script-based stage of the three temporal perspectives. One way to do this is to confront the animal with unfamiliar scenarios from several domains, where the existence of a script or other innate or learned response tendencies is unlikely (Suddendorf and Corballis 2010, 296). Additionally, anticipatory planning tasks have been demanded to only contain one-trial or one-shot performances, because this way it can be ruled out that the relevant behavior has been routinely acquired through repetition (ibid.).

Because semantic and script-based knowledge often seem to be implicitly used interchangeably, some argue that the successful performance in one-trial experiments proves that episodic rather than semantic future thinking has been at work. This is highly controversial, however, because one-trial learning in animals is actually poorly understood and occurs in other contexts besides episodic cognition as well (Osvath 2016, 10; but see Kabadayi and Osvath 2017).

Even more problematic is that the simple calculation 'if the anticipation of a future event is not based on scripts, it must necessarily result from episodic future thinking', may very well be too simplistic. In chapter 4.3 it was suggested with Szpunar (2010) that not every simulation of a specific personal future scenario necessarily involves episodic details. John may be able to simulate a neighbor's reaction to his invitation for dinner without the need to engage in mental space travel or mental time travel. Instead, he predicts that Mr. Miller will thank him very politely, but refuse the invitation (on the grounds that he cannot leave his dog unattended), without recalling a specific encounter with Mr. Miller, but based entirely on the semantic knowledge that his neighbor is very introverted and has a dog. This kind of simulation does not necessarily seem to depend on mental space travel or mental time travel. Neither, however, is it purely based on script knowledge, because it refers to a very specific personal future event. Also, John's knowledge of Mr. Miller's character trait is semantic, but results from several specific encounters with his neighbor. There may be many cases, therefore, where the contributions of semantic and episodic memory, respectively, are unclear. This means that just because the anticipation of a future event based on script-knowledge can be ruled out as, at least, very unlikely in one-trial experiments, it does not automatically allow any inferences with regard to the involved memory system.

To conclude, I believe that the spoon test with its proposed requirements (above all, the use of novel problems from a range of domains, the spatial-temporal separation between the preparatory behavior and the situation, to which the animal should have planned ahead, the relocation of the individual for testing and, optionally, the use of single trials) is a valid paradigm to investigate anticipatory planning abilities in animals. Beyond that, however, the spoon test *alone* (even if it involves only single trials) cannot inform us whether an animal planned ahead by help of semantic or episodic simulation, and, in the latter case, whether the simulation involved mental time or mental space travel. Fortunately, this does not mean that there is no way at all to find out whether animals may be capable of episodic simulation and whether, in doing so, they rely on mental space travel or even mental time travel, as will be argued in the next chapter.

5.1.2 *The Self-control Paradigm*

I have argued above that the spoon test paradigm provides a valid means to empirically investigate animal anticipatory planning. It cannot make any conclusive predictions, however, on whether the animal in question relied on semantic or episodic future thinking when anticipating her future needs. Therefore, another empirical paradigm is needed. This is in line with Osvaths' claim that there likely is no single test that can falsify all aspects of the Bischof-Köhler hypothesis (Osvath and Osvath 2008, 664).

With regard to the question whether an animal is indeed capable of mental time travel, it should be particularly promising to focus on the kind of experiments that have been laid out in chapter 4.2.2 and 4.2.4, aiming to understand whether and to what extent the animal in question can be said to have a grasp of the future. Only an animal capable of representing branching futures and of a minimal degree of hypothetical thinking does possess an essential requirement for mental time travel, to begin with. Thus far, chimpanzees have not been shown to solve the forked-tube task in not preparing for two mutually exclusive possibilities of how the future may unfold. While it definitely is too early to conclude from that that they are unable to represent branching futures, it opens an interesting avenue for future research. If further research should continue to find no evidence for an understanding of branching futures in great apes (and corvids), this should preclude them from the capability of mental time travel as well.

Whether an animal is capable of mental space travel or can 'merely' engage in semantic future thinking, might be more difficult or even impossible. To prove empirically that an animal solved a spoon test by help of mental space travel rather than semantic simulation is challenging, because "an undisputable sign of episodic pre-experience appears to be unattainable" (Osvath and Osvath 2008, 663). Nonetheless, some experimental designs make a simulation via mental space travel in contrast to a semantic simulation much more likely than others.

One experiment that does *not* provide the best conditions to provoke mental space travel on behalf of the tested individual may, in fact, be the spoon test paradigm. The reason for this is that one crucial element of spoon tests is that the current and the future motivational states of the subject must qualitatively differ or even be in contrast. John should go grocery-shopping now, even if he is currently sated. Analogously, an animal capable of anticipatory planning should prepare for her future desire for one food-type, even though she is currently sated on this type of food and currently prefers to eat another. In situations, in which the present and the future motivational states qualitatively differ, however, it is not easily comprehensible how a subject should be able pre-experience a desire for a certain food-type, while also experiencing a satiation on it. It should be more plausible to assume that both John and the animal in the example above do not episodically pre-experience their future desires, but rather semantically predict it (ibid.). (Of course, this is not to claim that episodic pre-experiencing can *never* occur in spoon test situations.)

If spoon tests might not provide ideal conditions to investigate mental space travel, which kind of experimental paradigms could be better equipped to do so?

Reconsidering the main functions of episodic mental space travel simulation, which have been described in chapter 4.3, situations, in which the individual struggles between two motivations, where one refers to the present and one to the future, should plausibly increase the likelihood of pre-experience via mental space travel. It should be promising, therefore, to lay focus on delayed-gratification and self-control tests, where the animal must reject an immediately available but smaller reward in anticipation of a larger reward. If one of the main functions of mental space travel is to allow the individual to pre-experience the positive feeling associated with the future outcome, then an animal with mental space travel capacities should be motivated to reject a presently available reward and rather wait for the larger one, for instance.

Mental space travel can occur in the absence of planning, however. Thus, self-control experiments with the aim to verify planning abilities need to ensure that the motivation to wait for the large reward really is a result of prospection, an aspect that is emphasized by Mathias and Helena Osvath as well: "A cue to a future event might eventually result in a drive state, for example a craving, that in turn evokes a motivation to plan, however, it is not the drive state per se that evokes the prospection, rather it is the other way around" (ibid. 664). Put differently, an animal must be able to anticipate the future, in the first place, in order to be capable to plan accordingly for it. Most importantly, "the stimuli related to the future event must be controlled for not having an associatively learned connection to the upcoming reward. If there does not exist an associatively learned link, then the potential cuing should be compatible with true planning for future needs" (ibid.).

Once it has been established that the animal in the self-control experiment actually is planning ahead, the question can be raised whether the animal is doing so via semantic or episodic future thinking. The self-control paradigm offers an operationalizable distinction between semantic and episodic prospection. Even though semantic future thinking is not be equated with script-based knowledge, as has been argued above, it first and foremost is concerned with *regularities*. John and Angie, for instance, can predict their neighbor's reaction and their first impression of the Christmas party, respectively, because they have acquired, through repeated experience, knowledge about certain generalities. The neighbor is introverted, but friendly, so he will most likely politely refuse John's invitation. There has always been a Christmas tree at the party, so there will most likely be one this year as well.

The episodic details in a simulation, in contrast, refer to the *particularities* of an upcoming event, where the individual likely does not yet have a general or abstract knowledge about what to expect. If John has met his neighbor only once before, he will probably not merely rely on semantic knowledge about neighbors in general, but episodically recall the incidence when predicting his neighbor's reaction. If Angie went to her company's Christmas party last year, she will very likely recall episodic details of this event when mentally constructing the future scenario. Thus, even though it remains to be true that it is extremely difficult, if not impossible, to discriminate whether someone predicts what is going to happen, based entirely on semantic information, or whether someone imagines oneself to be in the said situation while recalling episodic details from past experiences, episodic future thinking should at least be much more likely to occur in novel rather than familiar situations. Testing animals in scenarios and contexts, for which they should be expected to be unable to rely on semantic information, thus, can thus strengthen the hypothesis that the said animals are capable of episodic mental space travel simulation.

Osvath and Osvath (2008) propose to manipulate self-control experiments in such a way that the animal does not merely have to pick the familiar, functional tool over nonfunctional distractors (e.g., a food reward or a toy) in order to successfully solve a future task, but is unexpectedly forced to choose another functional tool among several items, of which most or even all are unfamiliar. This means, crucially, that the animal cannot rely on any kind of previously learned information regarding the new tool. If the animal succeeds in picking the right tool, even though she has never seen or used it before, this would plausibly indicate that the animal *imagines* to use the item in the familiar task, because she would have to mentally compare each item with regard to its functionality (for an empirical example of Osvath and Osvath's proposed modifications, see Osvath and Osvath 2008, which will be discussed and analyzed in detail in chapter 5.1.3.3).

To conclude, the modified self-control paradigm as suggested by Osvath and Osvath (2008) offers a promising opportunity to investigate episodic future thinking via mental space travel, even though *conclusive* behavioral evidence will very likely turn out to be unobtainable. Concerning animals' mental time travel capabilities, it will be particularly interesting to know whether future research will find evidence for some animals being capable of solving the branched-tube task or similar experiments.

5.1.3 *Empirical Studies*

This chapter does not aim to provide an exhaustive overview of the *status quo* of comparative prospective cognition research. Rather, few studies have been singled out, which are commonly agreed on making a particularly strong case for planning abilities or future thinking in animals. For an excellent overview of the current status of prospective cognition research in animals, see, for instance, Redshaw and Bulley (2018) or the slightly older, but very extensive overview by Cheke and Clayton (2010). My aim is to discuss some of the most well-known and highly reputed studies with regard to their explanatory powers concerning the prudential unity relations and, consequently, an individual's time-relative interest in continuing to live. In each case, I will proceed by asking the following questions: what does the study aim to show?, what are the results?, and finally: what does the study actually show with regard to the conceptual classifications that have been made throughout chapters 3 and 4?

The studies will not be discussed in the order of their dates of publication, but with regard to the similarities of their research questions or hypotheses. Furthermore, because studies have thus far almost exclusively focused on corvids and great apes, I will mainly limit myself to these two groups as well.

As Redshaw and Bulley (2018) point out, ecological studies are crucial for determining whether animals depend on their future thinking and planning abilities outside the laboratory and, therefore, play a decisive role in answering whether animals are merely *capable* of planning or whether they actually *engage* in planning in their natural habitats (see also Malanowski 2016). The problem with ecological studies on planning behavior is, however, that there are so many more alternative explanations available than for experimentally controlled studies. The strongest arguments in favor of animal planning commonly draw on experimental studies. As will become evident, even these controlled studies quite often generate a heated debate on how the results should be interpreted. I am particularly interested in the explanatory forces of those studies, which have been argued to provide *especially strong* evidence for planning abilities in animals; thus – even though ecological studies will undoubtedly add another important piece to the overall picture of future thinking and planning abilities in animals –, I will not consider ecological studies here.

5.1.3.1 Anticipatory Planning in Corvids

One of the best-received studies concerning future thinking in animals was conducted by **Sérgio Correia et al. (2007)**. The aim of the study was to directly refute Suddendorf and Corballis' (1997, 2007) Bischof-Köhler hypothesis, arguing that animals were incapable of anticipating and acting for a future need. Following Tulving's proposed requirements of the spoon test, Correia and colleagues introduced the *specific satiety paradigm*, with the purpose to manipulate the animals' current desire for a specific food type in order to make sure that they actually anticipated a future need that qualitatively differed from their current one.

Like many other animals, Western scrub jays (*Aphelocoma californica*) are prone to the specific-satiety effect, which means that, when being prefed with a particular food A, they become specifically sated on this food-type and develop a preference for an alternative food B. After ensuring that the birds did not have natural preference for one food-type, but showed roughly equal liking of both A (dog kibbles) and B (pine seeds), the authors divided the birds into two groups, where one group functioned as a control group. Correia et al. conducted 3 trials in total, with each trial consisting of 4 stages, in turn.

On stage 1, the birds could eat either dog kibbles or pine seeds (hereafter A) for 3 hours, which resulted in a specific satiety for the other food type (hereafter B). Afterwards, they were given the opportunity to eat and cache both food types for 10 minutes. Because of the specific-satiety effect, the birds cached more food of the type they now prioritized, that is, food type B. 30 minutes later, the birds received another prefeeding round, where one group received access to the same type of food A from stage 1 (Same group) and the other was given food of the other type B (Different group). Analogously to the first stage, this period lasted for 3 hours. In the final stage of each trial, then, the birds were allowed to recover the caches they had made previously on stage 2 (Correia et al. 2007, 857).

As the birds in the Different group had developed a specific satiety on food B on the third stage, it is plausible to assume that they were not satisfied when they recovered their caches, which contained more food B than A. In case of the Same group, in contrast, which had developed a priority for food B on the third stage, the recovered caches matched their preferences. The authors hypothesized, thus, that if scrub jays can anticipate their future needs and also understand that their behavior can influence the outcome, then the birds of the Different group should change their caching behavior on stage 2 on subsequent trials. Put differently, just as John, who just had lunch, goes grocery-shopping in anticipation of his future hunger, the scrub jays, sated on dog kibbles, should still cache kibbles on stage 2, if they are capable to know that they will develop a desire for this currently non-desired food later.

In line with these predictions, Correia et al. (2007) found that the birds of the Different, but not the Same group in fact switched their caching behaviors. Contra the Bischof-Köhler hypothesis, they concluded that the birds were seemingly capable of anticipating and acting in accordance with their future needs, even when these needs were in stark contrast to their current ones.[91]

Correia et al.'s conclusions have been criticized. Because the birds in the Different group did in fact not cache *more* food A, but *less* food B, this was argued to show that the birds did not in fact prepare for a future need, but only that the birds had quickly learned not to cache food that would turn out to be of diminished value. Rather than preparing for a future need, the birds simply reacted to the diminished value of B and switched to caching less of it (Suddendorf and Corballis 2008, e1; see also Shettleworth 2010b, 393).

However, specific satiety is rationally expected to result in a decrease in desire for the previously fed food and not an increase in desire for the other,

91 Correia et al.'s study can be read as a response to one specific kind of criticism that had been raised against a very similar study from Nicola Clayton's lab, conducted by Raby et al. (2007). The set-up was the following: after the scrub-jays had experienced that in one compartment they would be given food in the morning (the breakfast compartment) and in the other they were not (the no-breakfast compartment), they were given the opportunity to eat and cache food in the evening. The jays responded to this by caching more food in the no-breakfast compartment, seemingly preparing for a future need. As critics (and the authors themselves) pointed out, however, it could not be ruled out that the birds followed a simple heuristic in line with a Pavlovian account, for which no future thinking would be required. Thus, the birds may simply have avoided caching food in a place associated with plenty of it and rather cached food in the compartment that had been associated with a lack of food (Cheke and Clayton 2010, 923; Roberts and Feeney 2009, 275).

which makes this objection untenable. Furthermore, if the birds really behaved in accordance with an incentive learning account, they would have been expected to cache less of *both* food types, because the birds had been prefed on both kibbles and pine seeds in their first trial (Cheke and Clayton 2010, Clayton et al. 2008). Also, the birds continued to preferentially eat food B on stage 2 over the course of the experiment. Thus, their eating and their caching behavior were diametrically opposed. This makes it plausible to assume that the birds' caching-behavior really was in contrast to their present motivational state, strengthening the authors' conclusion that the birds planned for a future need (Correia et al. 2007, 857f).

Another objection, however, was taken serious by the researchers. Because the birds had only been tested for their future thinking abilities in the context of food-hoarding, many skeptics raised the question whether their seemingly planful behavior may not just indicate a clever adaptation of a domain-specific, hard-wired ability, rather than a domain-general cognitive ability such as anticipatory planning (Suddendorf and Corballis 2008, e2; Shettleworth 2010b, 393; McCormack and Hoerl 2011, 136). The reasoning behind this thought is that a food-hoarding system that can quickly adapt to changes in the environment and thus ensures the satisfaction of one of the animal's most basic needs in the future would be extremely beneficial for the individual and might have evolved as a hard-wired ability.

Directly addressing these concerns, **Can Kabadayi and Mathias Osvath (2017)** recently published a very well-received study on anticipatory planning skills of ravens (*Corvus corax*) in a non-caching context. The authors argue that in order to verify domain-general planning, animals need to be tested in tasks where they lack ecological or behavioral predispositions (Kabadayi and Osvath 2017). Here, only the first two experiments of the study will be discussed, because the other two have a slightly different research question, which will be reconsidered below.

The authors selected two different conditions: tool-use and bartering, that is, the exchange of a token for a reward. Because neither tool-use nor bartering constitute behavioral predispositions of the birds (ravens are not habitual tool-users), a positive finding would plausibly imply that ravens possess an underlying cognitive ability to plan for future needs, which they can flexibly use in multiple domains.

After having learned that only a specifically shaped stone, but none of the other available items (hereafter called distractors), could open an apparatus that contained a food reward, the birds were allowed to interact with the apparatus. As no tool was available, however, the birds could not open the apparatus, which was then removed in the birds' presence. One hour later, in a different compartment, the ravens were allowed to choose one of 4 items from a tray, which contained the right tool to open the apparatus and 3 non-functional distractors. 15 minutes (experiment 1) or 17 hours (experiment 2) after they had made their choice, the apparatus was re-mounted.

In the bartering condition, the proceeding was analogously to the one described above in the tool condition: the birds first learned to exchange a specific token (a blue bottle-cap) for an immediate food reward and were, on the next day, exposed to an experimenter, asking for the token, which the ravens did not possess then, but which they could choose from a tray 60 minutes later. 15 minutes (or 17 hours) later, an experimenter asked for the token.

These experiments tested whether ravens can anticipate and prepare for an event in the non-immediate future. As Kabadayi and Osvath state, "[t]o succeed they had to select a functional item in the absence of the problem, keep track of the item during the delay and later use it when presented with the problem at a different location" (Kabadayi and Osvath 2017, supplementary material, 5). As the authors infer from their study that ravens *plan* for events 15 minutes or even 17 hours ahead in the future, I take their hypothesis to be that if the ravens turned out to select, save and then use the correct object, this would sufficiently indicate the birds' planning abilities.

The results matched the authors' expectations. In detail, the results for the first experiment (15 minute-delay) were as follows: 2 birds selected the right tool on all 14 trials (and used it on 12 trials) and the other 2 birds chose it on 11 out of 14 trials (and used it on 11 and 8 trials, respectively). Importantly, None, the bird who used the tool on only 8 trials, invented a novel way to open the apparatus without the tool during the experiment and was, therefore, excluded from tool conditions on subsequent experiments. If those trials are not considered, the mean success rate for all subjects was 86 percent. In the bartering condition, 3 birds selected the right token on all 36 trials and 1 bird on 35 trials. On average, the tokens were traded for the reward in 77.6 percent[92] of cases.

In line with the requirements for anticipatory planning experiments, that have been laid out in chapters 5.1.1 and 5.1.2 (though see Osvath 2016), the authors also put special emphasis on first trial performances. The birds' high success rate in first trials seems to strengthen the authors' conclusion that the

92 This comparably lower success rate is argued to be due to the fact that in the first experiment, the birds could select and exchange 3 tokens in one trial. The high error rate "suggests that three food rewards were too much for the subjects and they preferred not to exchange all tokens for the reward in one trial (and perhaps even save the tokens for the future)" (Kabadayi and Osvath 2017, supp. mat., 10).

birds were in fact planning ahead and preparing for a future event, when they selected an item from the tray.

On the second experiment, where the delay between the selection of the tool and the chance to get the reward stretched to 17 hours, 2 individuals chose the functional tool on all 6 trials (and used it on 6 and 5 trials, respectively) and one individual chose and used it on 5 trials (mean success rate: 88.8 percent). In the bartering condition, the ravens even selected and exchanged the token on all 6 trials, with one bird trading the bottle cap in 5 out of 6 trials. This means that the 3 ravens succeeded in this task in 95.8 percent of the cases.

It is not entirely clear whether Kabadayi and Osvath (2017) conclude from these 2 experiments *alone* – the study comprises 4 experiments in total – that the ravens anticipated and planned for future motivational states. The authors considered the first experiment's aim to show that the ravens "prepare for a tool-use and a bartering event 15 minutes into the future" (ibid. supp. mat. 5) and the second to "test whether ravens can solve tasks involving long delays just as apes can" (ibid. supp. mat. 6). In spite of these relatively modest hypotheses, it seems fair to assume that the authors regard the above results as sufficient behavioral evidence for planning abilities in ravens (see also Boeckle and Clayton 2017). Below, I will consider the validity of this interpretation.

It is potentially problematic that Kabadayi and Osvath's (2017) study does not satisfy the first requirement of spoon tests, that is, that the birds' present and future motivational states do not qualitatively differ from one another. The authors explicitly state that the first stage of the experiment, where the birds are exposed to the apparatus and the experimenter, respectively, in absence of a functional tool or token, has the purpose to "create a possible incentive for later planning" (Kabadayi and Osvath 2017, 203). This supposedly means that the ravens should develop a desire for the relevant tool or token, which will enable access to the food reward. This desire, however, is not in conflict with a future motivational state of the birds, in contrast to Correia et al. (2007)'s specific satiety paradigm.

A possible response to this objection might be that there can be other methodological means to investigate anticipatory planning besides the spoon test. This may be true. Still, however, I will argue below that Kabadayi and Osvath's (2017) study does not provide sufficient evidence for anticipatory planning in ravens. I will do so by attempting to answer what the first two experiments actually show with regard to the conceptual clarifications that have been made throughout the last chapters.

With regard to the first-trial performances of the ravens, it should be relatively unsurprising that the birds picked the functional tool and token, respectively, because they had learned their function during training. Crucially, in doing so, the ravens did not have to imagine themselves opening the apparatus or trading in the token, which is to say that they did not have to engage in a simulation of a future scenario to succeed.

Compare this to Jane finding a chip for a trolley on her way to the supermarket. She might simply pick it up from the pavement in acknowledgment of the item's functionality. This, however, does not mean that she also *plans* to use the chip in a specific future event. If it is accepted that planning necessarily requires the anticipation of a future event and that it is not enough to merely appreciate an item's functionality or general purpose, then the fact that the birds picked the right item instead of a distractor might in fact not be very telling with regard to the animals' planning capabilities.

That the birds later also *used* the tool or token, when the chance came up in the shape of the apparatus or the experimenter, indicates an understanding of the functionality of the cached item and also the knowledge of its current location. This, however, can sufficiently be explained with temporal updating. (The tool condition will be used exemplarily to specify this, but it applies to the bartering condition as well.) When the birds were confronted with the apparatus again, they plausibly had the motivational state or desire to possess the correct tool. Due to their ability to continuously update their model of the world and objects' whereabouts, they knew where the tool was hidden (without necessarily also being able to remember putting it there a while ago), which enabled them to get the tool and open the apparatus. Note that if the temporal updating interpretation is correct, then it does not matter whether the delay between the caching and the using of the tool amounted to 15 minutes or 17 hours.

In an intermediate conclusion, the first two experiments of Kabadayi and Osvath's (2017) study do not provide unequivocal evidence for anticipatory planning in my view. I acknowledge, however, that the authors themselves may even agree with me on this, potentially arguing that only all four experiments in total provide conclusive evidence for the ravens' planning abilities, so that my argument should not be automatically read as an attack against Kabadayi and Osvath (2017). I will come back to the remaining two experiments below.

What remains to be clarified is whether the first two experiments are sufficient evidence for the ravens engaging in mental space travel. Above, it was argued that the birds, at the very least, understood that to open the apparatus, they needed to get the tool. This may plausibly indicate that they mentally represented this tool, but it is not sufficient to ascribe the birds with mental space travel, if mental space travel is defined as to involve a simulation of a sequence of events. Whether the ravens did in fact simulate using the tool to obtain the reward (which would point towards the mental space travel explanation) or whether they 'just' possessed a visual image of the required item (which many animals should be able to do in the absence of mental space travel), should, however, be very difficult to resolve unequivocally. To conclude, these first two experiments of Kabadayi and Osvath's (2017) study do not permit conclusive evidence for the ravens to engage in mental space travel.

The claims that have been made so far may be disputed on the basis that Kabadayi and Osvath (2017) actually orient themselves very closely to another study conducted by Redshaw and Suddendorf (2013), which has been argued to provide evidence for episodic future thinking in 4-year-old children (see also Boeckle and Clayton 2017). Directly comparing the ravens' and the children's first-trial performances, Kabadayi and Osvath argue that the ravens actually out-performed the 4-year-olds (Kabadayi and Osvath 2017, 203).

A closer look on both studies reveals, however, that there are several differences, which caution making a too direct comparison between the performances of the children and the ravens. Analogously to the experimental set-up in Kabadayi and Osvath's study, the children in Redshaw and Suddendorf's experiment needed to choose a tool (here, a triangular-shaped key) among several distractors (other keys). When the children chose their item, however, they had been *told* that they would be going back to the room with the box (which could be opened with the triangular-shaped key) in 15 minutes; that is, the children made their choice in expectation of a future event. The birds, in stark contrast, had no reason, prior to their first trial, to expect that they would gain access to the apparatus. It seems unjustified, therefore, to argue that the animals planned ahead to an event 15 minutes into the future (for a related point, see the critical review by Redshaw et al. 2017).

Furthermore, the children needed to *infer* that the triangular-shaped key would open the apparatus in the first room, because, importantly, they had never used it before. During training, they had learned that a square-shaped key opened a box with a square-shaped keyhole and that the square-shaped key did not open the box with a triangular-shaped keyhole. When being asked to choose one key among several keys later, the children had to infer, therefore, that the triangular-shaped key would provide access to the box. Thus, in contrast to the birds, who had been allowed to learn the tool's functionality, the children needed to go a step further than that, because the key did not yet have a functionality for them, just like the distractors. Thus, in selecting the correct key, it seems more plausible, compared to the ravens, that the children in fact simulated using the novel key to open the box, and, thus, actually engaged in mental space travel.

Importantly, this does not automatically allow the inference that the 4-year-olds also planned ahead into the future, and, thus, were capable of anticipatory planing. In fact, the results of Redshaw and Suddendorf (2013) strengthen the point that has been made above, that is, that this experimental set-up may not be ideal for an empirical investigation of anticipatory planning abilities. The reason for this is that half of all participating children selected the correct tool among the non-functional distractors, a result well above chance. Interestingly, however, only 50 percent of these successful children, whose behavior could be interpreted as to indicate future thinking abilities, explicitly verbally referred to the key's *future* utility, when explaining their choice.[93] For the remaining children, it cannot be ruled out, thus, that the sight of the triangular-shaped key merely 'triggered' them to pick it (because of their previous exposure to the box with the triangular-shaped keyhole) or that they had a vague idea of its functionality, without actually planning for a particular *future* event. Similarly, the ravens in Kabadayi and Osvath's (2017) study may have picked the correct items without actually engaging in anticipatory planning, as has been laid out above.

In spite of these negative conclusions with regard to Kabadayi and Osvath (2017), it has been argued that Correia et al. (2007) makes a strong argument for anticipatory planning abilities in Western scrub-jays, even though it remains possible until now that the seemingly planful caching behavior of the birds reflects a domain-specific, hard-wired adaptation of their food-hoarding system, rather than a domain-general capability like anticipatory planning. I would like to conclude this chapter with an analysis of what an acknowledgment of Correia et al.'s interpretation indicates with regard to their capability to engage in episodic future thinking. Thus, it will be assumed for the remainder of the chapter that scrub jays really are capable of anticipatory planning, that is, that they can anticipate a future desire and act in a way that ensures the satisfaction of this future motivational state.

Did the birds anticipate their future desire via semantic or episodic mental space travel simulation? On the one hand, as has been repeatedly mentioned throughout this thesis, the role semantic memory plays for future thinking is still controversial. In fact, it has been argued that a future mental state often can be predicted without the need to mentally project oneself into the future situation itself. Whether the birds pre-experienced a future desire or just predicted semantically that they will face a certain need, cannot be answered conclusively by help of this experiment, therefore.

On the other hand, episodic future thinking becomes much more probable in cases where the individual has only had very few relevant prior learning

93 100 percent of the children, who verbally referred to the key's future utility, also chose the correct key.

experiences and where the anticipated state in question refers to an experiential state. If mental space travel has the primarily function to bring the feel of a situation into the present in, for instance, allowing the individual to pre-feel a certain emotion associated with this situation, as has been argued in chapter 4.3, then those cases, in which the individual needs to anticipate her own emotional or motivational states in a situation (hunger, anger, appetite etc.) should be particularly well-suited to result in mental space travel. To conclude, the fact that the birds were seemingly able to anticipate their future motivational state is not conclusive evidence for or against the involved mode of future thinking, but a mental space travel explanation may indeed be more probable in this particular case.

5.1.3.2 Dissociation of Two Future Events in Corvids

The above mentioned study by Correia and colleagues (2007) found that Western scrub jays can, in contrast to the predictions of the Bischof-Köhler hypothesis, anticipate a future desire and act appropriately, even if this future desire is in contrast to the birds' present motivational state. Another study from Nicola Clayton's research group investigated whether jays are also capable to prepare for *two* distinct future desires they would experience one after the other. Prior to that, it had been argued that subsequent animal planning research should lay more focus on the ability to dissociate different points in the future (Roberts and Feeney 2009, Roberts 2012).

To do so, **Lucy Cheke and Nicola Clayton (2012)** relied, once again, on the specific satiety paradigm described above to control the current and future motivational states of the individuals. The 4 jays participating in this experiment were not Western scrub jays, but Eurasian jays (*Garrulus glandarius*), another food-caching corvid. The birds in the relevant planning experiment received 3 trials, where each trial consisted of 3 stages, in turn. The first trial familiarized the birds with the process of the experiment, so that trials 2 and 3 represented the actual test trials. In-between the stages, the birds had free access to their normal maintenance diet of kibbles, vegetables, fruit and seeds and water. The procedure of Cheke et al.'s (2012) experiment was rather complex, so I will only focus on the crucial elements here.

On stage 1 of the first trial, the birds were prefed their normal maintenance diet for 15 minutes. Afterwards, they were allowed to eat and cache food A (peanuts) and B (raisins[94]) in trays 1 and 2 (individuated by colored blocks) for another 15 minutes. As a previous control experiment had shown that

94 Some birds received pellets instead of raisins, depending on individual preferences. The general experimental procedure was the same, however.

preferences for either one of the two food types was not significant, the birds were expected to cache approximately equal amounts of peanuts and raisins into both trays, which were then removed.

On stage 2, the birds were allowed to feed on one of the two test foods (e.g., peanuts) in powdered or liquidized form (as to prevent attempts of caching) for 15 minutes. Afterwards, the food trays returned, containing the caches that the bird had made earlier on stage 1. The birds were allowed to retrieve and feed on their caches from tray 1 for 15 minutes. (Tray 2 had been masked with a transparent cover.)

Stage 3 occurred one day later. The procedure was exactly the same as in stage 2, except that the birds were now prefed the other food-type (i.e., raisins) and that this time, only tray 2 was accessible afterwards.

In order to manipulate the birds' current and future motivational states, the procedure of the first stage on subsequent trials now changed: rather than having free access to their normal food at the beginning of stage 1, the birds were prefed one of the two food types (e.g. peanuts). Based on the results of Correia et al. (2007) and in line with the specific satiety hypothesis, they should thus have a motivational state for the kind of food they had not been prefed, that is, raisins, when making their caches.

Yet, the birds had experienced that in the future, they would only gain access to the two trays one after the other. First, they would be allowed to retrieve caches from tray 1 (stage 2), afterwards they would be provided access to tray 2 (stage 3). Because they would also be prefed peanuts later on stage 2, the birds would first develop a desire for raisins. It would be better, consequently, if tray 1 contained more raisins than peanuts. On stage 3, in contrast, the birds would prefer their caches from tray 2 to contain mostly peanuts, before they would be prefed raisins beforehand.

This complex experimental procedure allowed Cheke et al. to hypothesize that if the birds understood that they would develop two desires in the future, that would qualitatively differ from one another, they should accordingly change their caching behavior on trials 2 and 3. In detail, if they were actually able to anticipate and dissociate their future desires for raisins on stage 2 and for peanuts on stage 3, they should preferentially cache raisins in tray 1 and peanuts in tray 2.

The results matched this hypothesis. Whereas the jays cached both peanuts and raisins on their first trial, they preferentially cached the kind of food in each tray they would desire later at retrieval. There was neither a general preference for one food over the other, nor did the birds reduce caching in total over the course of the experiment. As had been the case in Correia et al.'s (2007) study, the birds' caching differed from their eating behavior, indicating that they were indeed acting in accordance with their current motivational state when eating, but not when caching (Cheke and Clayton 2012, 172ff).[95]

Of special interest for this thesis' research question now is, of course, whether the birds based their decisions on an understanding of the temporal relationship between the two future events x and y, or whether they just associated one tray with one distinct event, regardless of when in the future it would occur. If the jays knew that one event would precede the other, they should also understand that one event would occur *sooner* than the other. To investigate this, **James Thom and Nicola Clayton (2014)** conducted an intertemporal choice study, which will be presented as follows.

8 Western scrub-jays (*Aphelocoma californica*) participated in the first experiment and received 8 trials, with each trial consisting of a 3-stage-procedure, carried out across 3 days. On day 1, the birds were deprived of food for 2 hours and were then given the opportunity to cache mealworms across 3 trays, individuated through colored blocks. Each day at the same time, the birds received one tray and could retrieve the caches they had previously made. Tray 1, therefore, returned 1 hour later after caching, tray 2 returned 25 hours later and tray 3 returned 49 hours later. Between the recovery periods, the birds had free access to their maintenance diet. The authors hypothesized that if the birds were sensitive to the delays associated with the trays, they should be expected to preferentially cache food in tray 1, associated with a shorter delay. Strikingly, however, the birds did not change their caching behavior over the course of the experiment. Thus, no evidence of a preference for earlier recovery could be found.

To rule out the possibility that the delays between caching and recovery had simply been too long for the birds to develop any preferences for one delay over another, the authors conducted a second experiment. Furthermore, this experiment tested whether the birds might perhaps not so much care about shorter or longer delays between caching and recovery, but instead about the amount of time their caches remained available for recovery.

95 One Eurasian jay called Hunter did not consistently switch to the expected caching strategy. On the last trial, Hunter's preferences seemed to turn into the opposite in that she, contrary to her performance on the previous trial and also in contrast to the other three birds' behavior, preferentially cached peanuts in tray 1 and raisins in tray 2. The authors tentatively raise the possibility that Hunter may have developed another strategy over the course of the experiment: "Instead of adapting her caching to account for food consumed immediately before cache-retrieval, she may have decided *not to eat* before cache-retrieval. This would be a valid strategy, particularly if she preferred whole to powdered peanuts" (Cheke et al. 2012, 174).

Thus, the birds were allowed to choose between three options. First, they could cache in tray 1, which would be available for recovery 7 minutes later and which would remain in the cage until the end of the trial 2 days later. Alternatively, they could cache food in tray 2 with a 26-hour delay between caching and recovery, where the tray would be taken away after 100 minutes. As a final option, the birds could cache in tray 3 with a 7-minute delay between caching and recovery, but a recovery-period of only 100 minutes. Again, however, the birds appeared indifferent to the three conditions, suggesting, therefore, that neither of the two tested factors (the amount of time they had to wait for their reward and the amount of time, in which the trays would be accessible) did influence their caching behavior.

Thom and Clayton (2014) laid out several plausible interpretations of their null-results. For one thing, the birds may simply have failed to associate the trays with the particular delays, either because of insufficient learning experience – which the authors, however, considered unlikely given prior experimental set-ups (Raby et al. 2007, Correia et al. 2007) – or because of their inability to dissociate future delays. Alternatively, the birds may have been aware of, but indifferent to the different delays. Given the benefits a (conscious) delay-sensitivity has with regard to cache theft, however, this finding should be very surprising, to say the least. Thom and Clayton (2014) do not rule it out, however, and regard this as a further plausible interpretation of the scrub jays' lack of temporal preference, therefore.

Correia et al. (2007) had found that jays can anticipate events up to 3.5 hours into the future, so maybe the delays used in this study (up to 26 hours) were located too far in the future so that the birds "might rely on the simpler, inflexible caching drive alone" (Thom and Clayton 2014, 178) rather than on their anticipatory planning abilities, when the relevant events were beyond their temporal horizons. To test this hypothesis, more intertemporal-choice experiments with jays (or other corvids) would be needed. If future experiments still reported null-results, a general incapability to dissociate future delays would become likely. Similarly, Thom and Clayton conclude:

> Future studies could clarify the results of this study by assessing the scrub-jays' capacity to discriminate between different time points in the future. Investigation of scrub-jay intertemporal choice in a non-caching context would complement the caching studies, and would be more directly comparable with previous work using other species. Together, this work would clarify our understanding of scrub-jay intertemporal choice, and could shed some light on the adaptive function of complex cognition in caching. (Thom and Clayton 2014, 178)

The impressive findings of Cheke and Clayton (2012), on the one hand, and Thom and Clayton's (2014) null-results, on the other hand, initially pose a challenge to the attempt to unify them into one plausible account of jays' planning abilities. Above, it was argued that Correia et al.'s study (2007) provides sufficient evidence for anticipatory planning abilities in jays, at least, if the problem of domain-specificity is disregarded here. Cheke and Clayton's (2012) results on Eurasian jays strengthen this conclusion and show that the birds can even plan for two temporally distinct future events, associated with two different motivational states.

Yet, Thom and Clayton's (2014) results strongly suggest that in doing so, the birds may not represent the intertemporal relationship between those events, that is, they may be able to discriminate two future events from one another, but not be able to know which one will temporally precede the other. This could be interpreted to relate to McCormack's intermediate stage of the development of temporal thought (chapter 4.4.3). Thus, the birds might have a grasp of the future as being alterable and be able to act appropriately to shape a future outcome, but not represent the future in a linear temporal framework.

If this, admittedly very speculative, stipulation will prove to be correct, then one hypothesis following from it would be that jays do not possess event-independent thinking abilities and will, thus, not be able to engage in complex hypothetical thinking. Because hypothetical thinking at the very least requires an ability to represent branching futures, a first way to make progress towards an answer here could be to empirically test jays in branched-tube experiments as described in chapter 4.2.2. (Naturally, the test would have to be modified to the specific physiological conditions of the birds in order to enable them to prepare for two mutually exclusive possibilities without hands, but this certainly is no insuperable obstacle.)

Whether the birds in Cheke and Clayton (2012) pre-experienced the two future desires via mental space travel or whether they just anticipated their future needs by use of semantic future thinking cannot be answered conclusively in this experimental set-up, as has been argued above in discussing Correia et al. (2007), but mental space travel may, again, be the more plausible explanation. The primarily function of mental space travel has been argued to be pre-experiential and should be particularly beneficial in those cases, where the individual has not yet have enough prior experience as to possess an abstract knowledge of certain preferences, but where it is more probable that her decision is also based on the retrieval of episodic elements.

As has been argued above, mental space travel, however, can also 'trigger' certain behaviors and be embedded in an associative learning mechanism rather than anticipatory planning. Notably, this is also explicitly emphasized by Cheke and Clayton (2012). In line with Boyer (2008), they argue that a mental representation could "give an action's (temporally distant) consequences emotional salience in the present and act as a counter-motivation against current desires. This process would be entirely outside cognitive control as it would be triggered without deliberate construction" (Cheke and Clayton 2012, 174). Similarly, the so-called *mnemonic associative theory* argues:

> [R]ather than previous outcomes being re-experienced at the time of action, previous actions are re-experienced at the time of the outcome, [which] leads to the formation of an association between action and outcome which drives drives future behavior in similar contexts. Thus, 'future-oriented' action is performed in response to previously formed associations rather than from cognitive decision-making in response to a retrieved memory. (Cheke and Clayton 2012, 174)

Essentially, both interpretations raise the possibility that when an individual remembers an event at the time of another event, this may lead to an association between these two events and could therefore result in long-delay associative learning (ibid.). According to the authors, this implies, somehow paradoxically, that some animals may be able to experience their future motivational states and simultaneously remain stuck in time in the sense that the Bischof-Köhler hypothesis argues them to be:

> [I]f Boyer is correct and [mental space travel] allows future motivational states to be experienced in the present, then [mental space travel] may have developed in some animals precisely because the [Bischof-Köhler hypothesis] is correct. If an individual can only act on a current motivational state, then the only way for them to be prospective is by changing that motivational state, rather than disengaging from it. (ibid.)

If this turned out to be correct, the jays' behavior should no longer be interpreted as anticipatory planning. Anticipatory planning, it has been argued, necessarily involves an anticipation of a future mental state, that is, the individual must at least understand that this anticipated state refers to a non-current state.

On the one hand, Cheke and Clayton's conclusions may be too pessimistic in light of their own empirical results, however. The authors, as well as Correia (2007), had observed that the birds changed their caching but not their eating behavior over the course of the trials. Thus, even if it may be true that mental space travel enabled the birds to pre-experience a certain motivational state, they still were seemingly capable of separating their current motivational state for caching from their current motivational state for eating.

On the other hand, the authors caution that caching birds like jays could possess "semi-independent motivational systems for eating and caching" (Cheke and Clayton 2012, 174), which would make such a simultaneous existence of two distinct motivational states in caching birds possible. Crucially, such an explanation would not require any planning capabilities on behalf of the birds.

This uncertainness emphasizes the need to investigate the domain-generality of planning abilities in caching birds: if jays or other corvids were shown to be capable of anticipatory planning in other domains besides caching as well, this should make a non-associative learning account much more plausible. For this reason, I will return to the last two experiments of the recently conducted study by **Kabadayi and Osvath (2017)**, which hypothesized that the ravens take delays into account and, thus, prefer events with a shorter waiting period over those with longer delays.

First, the authors were interested in whether the ravens would exert self-control when having to choose between a small, but immediately available reward (dog kibbles) and the tool or token, which would provide access to a highly valued food reward, available 15 minutes later. As had been the case in the previous two experiments, the ravens were tested in 2 conditions. In the tool condition, the birds could select a functional tool from a tray, which they could use later to open an apparatus containing a reward. In the bartering condition, the tray contained a blue bottle cap, which could be traded for a reward with an experimenter later. In each trial, the ravens were thus presented with a tray, which contained the functional tool or token, the distractors and an immediate reward. 3 ravens participated in the tool condition (14 trials) and 4 in the bartering condition (14 trials).

The birds selected the tool and the token, respectively, in about 73 percent in each condition, even though they did not always use the items accordingly later. In those cases, where the ravens did *not* take the functional item, they selected the smaller reward. In a control condition, when no functional item was available, the subjects took the small reward in 100 percent of the cases.

Finally, Kabadayi and Osvath aimed to investigate whether the ravens would take different delays into account when making their decision. To do so, the ravens in this experiment were led past the apparatus or the trading experimenter to provoke a desire for the tool or token and to communicate the information that apparatus and experimenter were within reach this time. The birds were then allowed to pick an item from the tray, which contained the same items as in the experiment before. In contrast to the previous experiment, where the birds had had to wait for 15 minutes before gaining access to the large reward, the food reward was quickly accessible this time, therefore. The authors suggested that the ravens should select the relevant item over the smaller reward even more often than they had done in the previous experiment.

Indeed, in both conditions, the ravens selected (and used) the tool or token in 100 percent of trials, representing a significant increase compared to the results of the prior experiment (73 percent). It appears, therefore, that when the delay between item-selection and recovery was rather long, that is, 15 minutes, the birds did not always regard the reward worth the wait, whereas they reliably picked the tool or token from the tray, when the delay was very short.

According to Kabadayi and Osvath, their study indicates that the ravens "made decisions for future events by inferring temporal distances to these events, which shows that the item is not selected because of an intrinsic value but because of its relation to a future" (ibid. 204) and that the birds "consider temporal distance to future events" (ibid. 202). In a similar fashion, Boeckle and Clayton argue in their review of the study: "The results [...] show that ravens take temporal distance between item choice and reward into account, exercise self-control, and make decisions for predicted futures rather than arbitrary ones" (Boeckle and Clayton 2017, 127). The underlying assumption seems to be, therefore, that if the birds do not base their decision on an associatively learned mechanism of attributing an inherent value to a tool, they must necessarily be taking temporal delays into account. It will be argued below that the birds' behavior may be interpreted alternatively, however, without having to assume this kind of temporal understanding.

By help of control conditions, the authors satisfactorily show that the birds did not base their choice simply on the inherent value of the functional item, that is, their selection did not simply represent an associatively learned behavior. Put differently, the birds did not just pick the tool or token, because they had learned to associate it with a reward but without understanding its functionality or future utility. It might very well be the case that in preferring the item over the immediately available reward, they knew what exactly the item 'stood for', that is, an even larger food reward. Maybe the birds even mentally visualized the large reward when making their decision. Importantly, however, this is not sufficient for mental space travel, because mental space travel requires a mental representation of a *sequence* of events rather than just a visualization of an object. But even if it is assumed that the birds did engage in mental space travel in, for instance, imagining themselves opening the apparatus with the tool and eating the food reward, this does not allow the inference that they also represented the 15-minute delay between selection and reward, because their decisions could simply be based on a preference for a large (mentally represented) over a small (currently perceived) reward. Let me clarify this claim in the following.

At first glance, the fact that the birds selected the tool or token significantly more often over the small reward, if the large reward had been available more quickly (experiment 4) than in those cases where the large reward had only been available in 15 minutes (experiment 3), seems to show that the ravens did indeed take the different delays into account, indicating an understanding of when in the future an event will occur. I will suggest that the ravens' behavior is alternatively explainable via temporal updating.

In experiment 3, the birds made their choice in the absence of the apparatus or the experimenter, whereas, in experiment 4, they made their selection with the knowledge that the apparatus or the trading partner already were accessible. I tentatively stipulate that this difference could have affected the birds' motivational states, explaining, in turn, the different results between both experiments.

In the third experiment, the birds were confronted with two tempting options. They could either take the small, but immediately available reward, or they could choose the functional item, hoping that the possibility to obtain the larger reward would arrive. Importantly, the birds could not be one hundred percent sure, whether and when this would actually be the case, however, so that some individuals might have preferred the small, but safe reward over the larger, risky one. Furthermore, the experiment requires some self-control from the birds, which may have led some weak-willed individuals to preferentially take the small reward, as Kabadayi and Osvath (2017) argue in a similar fashion.

In the fourth experiment, in contrast, the ravens walked past the apparatus or the experimenter, probably aware, therefore, that the only thing they now needed in order to obtain the large reward was the tool or token. They knew this, crucially, *before* making their choice. Knowing that the large food reward is accessible right away should make it much easier to choose the functional item over the small reward than in cases, in which the large reward remains only a vague possibility. Thus, when selecting an item from the tray, the birds plausibly knew for sure in experiment 4 – but not in experiment 3 – that they would get the reward, which may be sufficient to explain their behaviors. Significantly, this explanation does not make any claims as to the birds' reasoning about the involved delays. Instead, the ravens' behavior may be explained without the need to assume any anticipatory planning abilities, nor the ability to dissociate two different delays.

Most importantly, however, is the following objection. Even if it turned out in future experiments that corvids indeed exercise better self-control in situations with shorter rather than longer delays, this would not be a counter-example against Thom and Clayton's (2014) proposal that corvids may be unable to represent intertemporal relationships. A possible problem in making the comparison here could be that the jays in Thom and Clayton's study needed to decide between two delays of different lengths, whereas the ravens in Kabadayi and Osvath's self-control experiments merely had to choose between a delayed reward or an immediately available reward. To make sure that ravens can dissociate two future events with regard to their intertemporal relation, they would be required to choose between two different delays, as in Thom and Clayton's study, but not between delay and no delay, as in Kabadayi and Osvath's set-up.

In summary, Correia (2007) as well as Cheke and Clayton (2012), using the specific satiety paradigm to manipulate the subjects' present and future motivational states, provide the most compelling evidence for corvids' anticipatory planning abilities. Anticipatory planning is not restricted to situations, where the future motivational state of an individual qualitatively differs from the current one, but empirical investigations of anticipatory planning abilities of animals should confine themselves to those situations, because otherwise alternative explanations cannot be ruled out, as has been illustrated.

This interpretation is facing two challenges, however. For one thing, the birds have only been tested in caching-specific contexts, which raises the possibility that the birds' impressive-looking behavior could merely be due to a fast-working adaptation-mechanism. Kabadayi and Osvath (2017), attempting to investigate corvid planning in other domains, are unable to conclusively provide evidence for future thinking (and anticipatory planning), as has been argued above.

Furthermore, Cheke and Clayton (2012) themselves suggested in their extensive and well-thought-out conclusion that the jays may not have anticipated their own future motivational states; rather, their behavior could reflect an associatively learned behavior. In fact, it seems very difficult, if not impossible, to empirically verify whether a certain behavior results from a conscious anticipation of a future motivational state or was caused by the formation of an unconscious association between outcome and action. The first possibility should become much more probable, however, if corvids could be shown to execute seemingly planning behavior not only in caching-contexts, but other domains as well.

Concerning the corvids' mental space travel abilities, self-control experiments should be suitable means for measurement, in principle, because self-control represents an essential function of mental space travel. Mental space travel is especially beneficial, when the individual cannot depend on previous experiences or general information, that is, in novel or unfamiliar situations. Again, however, the role that semantic memory plays for future thinking has been a topic for hot debates for some time now, even further complicated through the fact that the distinction between semantic and episodic future thinking is far from consensual. It may be particularly difficult, thus, to provide *conclusive* behavioral evidence for episodic mental space travel simulation in animals.

It should be kept in mind that even if corvids turn out to have anticipatory planning abilities, this does not automatically mean that they are also capable of mental time travel. As a first step towards a verification of mental time travel abilities, temporal-causal reasoning abilities needed to be empirically investigated. A first, easy step into the right direction would be to test corvids in branched-tube tasks (or similar set-ups). Another strain of evidence for possible temporal-causal reasoning abilities in animals may potentially come from strategic planning experiments, which will be discussed in chapter 5.2.

5.1.3.3 Anticipatory Planning in Great Apes

One year after Correia and colleagues had published their findings on anticipatory planning behavior in Western scrub jays, **Mathias and Helena Osvath (2008)** presented their study with great apes. Until now, their thorough and well-controlled study is commonly agreed on making one of the best arguments for anticipatory planning and episodic future thinking abilities in great apes. I agree with this assessment, as will become evident throughout this chapter.

The authors conducted 4 experiments in total, where the last 2 served as controls for alternative explanations besides anticipatory planning and episodic future thinking. In detail, they had the purpose to make associative learning or semantic prospection explanations extremely unlikely. In all 4 experiments, 2 chimpanzees (*Pan troglodytes*) and 1 orangutan (*Pongo abelii*) participated.

After demonstrating how a transparent plastic hose could be used as a straw to suck up fruit soup from a box, the animals were allowed once to use the tool to obtain the reward themselves. In the first experiment, the apes were then allowed to choose one out of 4 items from a tray in another room, from which the apparatus was not visible. The tray contained the plastic hose and 3 non-functional distractor objects. One chimpanzee and the orangutan selected the tool on 100 percent and the other chimpanzee on 13 out of 14 trials. After they had made their choice, the animals returned to their daily enclosure, which they shared with other individuals. 70 minutes later, the apes got access to the reward room with the box. The subjects brought and used the tool in the majority of cases.[96]

These findings indicate that the animals were able to select and save a tool for future use, even when the reward site was not visible from the selection site. Moreover, an associatively learned connection between tool and reward is unlikely, because the subjects did only once, during training, gain an immediate reward after selecting the tool; afterwards, they always had to wait for at least 1 hour to obtain it. Moreover, they were not always rewarded after having chosen the correct tool, because of frequent tool losses during the delay, making the possibility of an associatively learned connection even more unlikely. Still, it cannot be conclusively ruled out, as the authors admit, so this concern was directly addressed in experiments 3 and 4, which will be presented later.

What conclusions does the above experiment allow regarding the great apes' planning abilities? Sharing their waiting area with the other members of their group "presumably added a cognitive load to the subjects, not only when it comes to keeping the tool from getting lost, but also in the context of all the attention grabbing episodes that occur in ape enclosures in zoos" (ibid. 667). The authors conclude, thus, that "it is unlikely that the desire for fruit soup was the predominant drive state during the delay" (ibid.).

Even if this claim is accepted, it does not rule out the possibility that the apes chose the tool simply in acknowledging its general function, rather than in anticipation of a specific future event, and then took it with them to the enclosure. Later, when the apparatus was mounted in the other room, the desire to retrieve the fruit soup emerged, so the apes picked up the tool and went to the reward room. Thus, the apes' behavior could be explainable via temporal updating rather than anticipatory planning (see Suddendorf et al. 2009, 752 for a similar criticism). It should be noted, however, that even though the first experiment by itself cannot rule out non-planning explanations, the 4 experiments in total make a strong argument for the authors' interpretation. I will come to the next experiment.

The second experiment tested the apes' ability to reject an immediate, highly valued reward (a grape) in favor of the hose, which would provide access to an even larger food reward (fruit soup) in the future. The procedure was

96 When they did not succeed in obtaining the reward, they had either lost or damaged the tool during the waiting period, so that it was not functional anymore. One chimpanzee's infant repeatedly misplaced the tools of her mother and the other chimpanzee, whereas the participating orangutan made the hose either nonfunctional, due to carrying it in his mouth, or inaccessible, dropping it outside the enclosure (Osvath and Osvath 2008, 667).

similar to the first experiment, with the exception that the tray now additionally contained a grape. The subjects chose the tool in 8 (chimpanzee Linda), 11 (chimpanzee Maria Magdalena) and 9 (orangutan Naong) out of 14 cases, selecting the fruit on all other trials. They also brought and used the tool in the majority of trials. When they did not, this was mainly resulting from tool loss due to other apes stealing or misplacing it.

According to Osvath and Osvath, these results suggest that great apes perform at least on par with, if not better, than human adult participants in comparable self-control experiments (Osvath and Osvath 2008, 668). Because the subjects in the experiment opted for the tool rather than the grape, Osvath and Osvath concluded that "great apes are able to make choices favoring future needs in direct competition with immediate drive states" (ibid.).

Against this interpretation it could be argued that when the apes selected the tool, they did so in grasping the general high utility and functionality of the tool, which sufficiently motivated them to favor it over the immediately available reward. This would only require an understanding of what the tool represented, but it would not necessitate the animals to simulate using it in a future scenario.

To rule this kind of explanation out, Osvath and Osvath conducted a clever control, which combined the previous 2 experiments. Just like in the first experiment, the apes were allowed to take one of 4 items from a tray, which contained the hose and 3 nonfunctional distractors. As expected, the subjects chose the functional tool. Immediately afterwards, the apes were again allowed to select an item from a tray, which now contained a copy of the functional tool, 3 distractors and a grape, similar to the procedure in experiment 2. Osvath and Osvath hypothesized that if the apes had merely associated the tool with the larger reward throughout the study, then they should choose the tool again in the second round. If, however, they understood that the tool had the function to obtain a reward in a particular future scenario, they should choose the fruit.[97] The authors found that all 3 individuals selected the fruit on all 14 trials, when they already possessed a copy of the tool.

This finding indicates that the apes knew that that the tool's function was independent from how many tools they held in their possession. Thus, the value of the tool did not remain the same for the animals over the course of the

97 As Osvath and Osvath point out, it would not necessarily be suboptimal for the animals to select the tool on the second pairing again, at least once in a while, because of the high risk of losing it to other group members during the delay. In prior experiments, loss of the tool had occurred in 20 percent per individual, on average. That the apes consistently chose the grape over the copy of the functional tool seems to be best explained by their strong desire for this particular type of food (Osvath and Osvath 2008, 669).

study, independent of the situation, which excludes an associatively learned connection between tool and reward. The question is, however, whether the apes necessarily needed to engage in *future thinking* in this experiment, as the authors suggested. Maybe the apes simply no longer had a desire for the functional tool, once they held it in their possession, in which case temporal updating would sufficiently explain the apes' behavior, without having to assume any anticipation of future states or desires. Suddendorf et al. (2009, 752) make a similar claim. This explanation, however, becomes very unlikely in the face of the study's last experiment.

The procedure of the final experiment was similar to the first one, with the exception that over the course of the experiment, only one of the 4 items remained the same, namely a non-hollow bamboo stick, which the apes were familiar with. Thus, in each trial, the other 3 items were completely novel to the apes. In detail, the tray in each trial contained 1 novel, functional tool to suck up fruit soup from the apparatus, 2 novel distractors and the familiar bamboo stick. The inclusion of this last item was of particular interest, because it had previously been used by the apes to extract honey and was, therefore, certainly associated with having a high utility.

The results of the experiment were as follows. The two chimpanzees selected and used the correct, functional item on 10 and 9 out of 12 trials, and the orangutan on 11 out of 12. Because they did not attempt to use a wrongly chosen item later to extract the fruit soup, the apes seemingly were well-aware of not being in possession of the functional tool in these cases.

The 12 functional tools were very dissimilar from the transparent plastic hose, that had been used in the task so far, and, importantly, sometimes also required the subject to take a different bodily position in order to obtain the reward (Osvath and Osvath 2008, fig. 1). Importantly, the apes only had visual access to the tools, that is, they could not touch the items before making their choices. This means, that in each selection situation, the apes were required to make a unique comparison between the features of the items and the reward situation in order to find out which of the offered objects could be appropriately used as a straw. As these tools vastly differed from the previously used plastic hose and also from one another, the application of a fixed rule or heuristic could almost certainly be ruled out. A heuristic such as '*Always choose the hollow object*', for instance, would not reliably have led to success in all tasks, because several trials included additional hollow objects, which could not be used as a straw. If the apes simply followed the above rule, therefore, they would have failed more often than they actually did. It seems plausible to assume, Osvath and Osvath (2008) argue, that the apes really did engage in anticipatory planning.

Even though the above study by Osvath and Osvath (2008) has been criticized from various sides, especially by Thomas Suddendorf (Suddendorf et al. 2009, Suddendorf and Corballis 2010; see Osvath 2010 for a reply), it has commonly been regarded as making a strong case in favor of anticipatory planning and episodic future thinking abilities in great apes. In the following paragraphs, some major objections will be addressed and replied to.

Osvath and Osvath deliberately decided against using the specific-satiety paradigm, because they were explicitly interested in the great apes' self-control abilities. Consequently, the present and future motivational states of the animals did not qualitatively differ from one another throughout the study, meaning that the study failed one important requirement of the spoon test paradigm.

One consequence from this is that even after the third experiment, there remain doubts with regard to the question whether the apes really did anticipate a specific desire or need they would have in the future, or whether their behavior is merely based on a broad semantic understanding of the tool's general function and utility. Even though Osvath and Osvath (2008) seem to hold that even the latter would at least indicate the great apes' semantic future thinking abilities, this can be questioned (see also Redshaw and Bulley, 2018).

As has been argued in chapter 4.4.3.4, one criteria for future thinking and anticipatory planning, irrespective of it being semantic or episodic in nature, is that the individual must have a grasp of the future that goes beyond an understanding of time in terms of aspectual notions. That means that the individual must at least have a sense of the future in terms of its alterability, rather than in terms of regularities and recurring events. Indeed, it has been argued that Osvath and Osvath's (2008) findings can be explained without ascribing any kind of planning abilities to the apes. I will argue that in light of all four experiments of the study, these objections do not hold, however.

Against the planning explanation one could argue that the apes might have quickly learned to expect the arrival of the fruit soup apparatus, which would explain their selection of the tool in the first experiment and also why they favored the tool over the small reward in experiment 2. Concerning the third experiment, this alternative explanation would additionally require the apes to understand that the general utility of the tool did not maximize with quantity, because they did not select the tool again in the second presentation of the tray, but this could still be well in accordance of the acknowledgment of the tool's *general* function without necessarily having to presuppose any anticipatory planning abilities. I take Sara Shettleworth's remarks on Osvath and Osvath's study to go into a similar direction (Shettleworth 2010b, 392). Slightly differently, Redshaw and Bulley (2018) propose that the sight of the novel, but appropriate tool may simply have 'triggered' a rigid mental representation of the reward, allowing the apes to make the correct choice without any reference to a *specific* future event, in which the apparatus would become accessible again.

In contrast to Shettleworth (2010b) and Redshaw and Bulley (2018), however, I take Osvath and Osvath's (2008) fourth experiment to make a rather strong claim in favor of the authors' conclusions that the apes did in fact base their selection of the novel functional tool on an (episodic) anticipation of a specific future scenario. The most plausible alternative explanation, that is, that the apes were somehow cued in selecting the correct item in every trial without actually anticipating the future scenario, is rather unlikely in light of the very different properties of the objects, I believe.

What is more, the cuing of the novel unfamiliar tool would have needed to be stronger than the cued mental representation of honey, which was plausibly simultaneously triggered by the sight of the bamboo stick. Even if it is assumed that the animals had been put in a predictive state with regard to the extraction of fruit soup rather than honey, it seems questionable that a totally unfamiliar item could outcompete a familiar tool in this way. The animals clearly understood that they were required to choose a tool for the extraction of fruit soup rather than honey and also that the only familiar tool in the tray, the non-hollow bamboo stick, was non-functional for this task.

If the apes had merely expected the retraction of fruit soup from the apparatus in terms of a recurring regularity, but not capable of acknowledging this event as potentially modifiable, then they would have been expected to become simply frustrated in the face of the presented items, knowing that the tray did not contain the one functional tool, they had been acquainted with. If, in contrast, the sight of the presented items caused them to *simulate* the function of the items in the specific task, it seems plausible that they did not only predict the future event of the arrival of the apparatus as a static, yet-to-come regularity, but as a potentially alterable event. This, in turn, would indicate a grasp of the future as being alterable in terms of McCormack's second of the three temporal perspectives.

The apes, I will assume, really did engage in anticipatory planning in the above experiments. But did they also simulate the future scenario episodically, as the authors suggest? Osvath (2016) himself cautions that even the use of single trials does not provide any conclusive evidence for episodic future thinking, because it is uncertain, whether this one trial, in which the animals become acquainted with the task, might be enough for semantic learning to occur. It might be questioned, therefore, whether Osvath and Osvath (2008) provide sufficient evidence for episodic future thinking in great apes.

It has been argued above that it might be impossible, or at least very difficult, to provide conclusive behavioral evidence for or against a mental space travel simulation explanation. However, if it is accepted that in selecting the correct, but unfamiliar item, the apes were seemingly able to mentally try out the use of several objects beforehand, then it seems indeed plausible to assume that they in fact episodically simulated the future scenario in mentally pre-experiencing the function and use of the objects (ibid. 671). This strengthens Osvath and Osvath's conclusion that the apes planned ahead with the help of episodic mental space travel simulation. Their study, therefore, provides strong empirical evidence for the great apes' capability to plan for a future event and engage in mental space travel. (Note, however, that this does not allow the conclusion that chimpanzees and orangutans are also capable of event-independent thinking and mental time travel.)

In sum, Osvath and Osvath (2008) make a strong argument in favor of anticipatory planning abilities and also of episodic mental space travel simulation of the two chimpanzees and the orangutan. Moreover, apes have been shown to use this ability not only in the domain of tool-use but token-exchange as well, which has been suggested to provide evidence for anticipatory planning being a domain-general capacity in these animals, even though the relevant studies have been disputed (for an overview of relevant studies and alternative explanations see Redshaw and Bulley 2018). Studies with monkeys, in contrast, so far suggest that they are unable to succeed in similar tasks.

5.1.3.4 Affective Forecasting in an Orangutan

Another study from Mathias Osvath's research group provides further evidence for episodic future thinking abilities in great apes, even though it does not directly assess planning. **Gabriela-Alina Sauciuc and colleagues (2016)** were interested in whether an orangutan would be able predict and pre-experience the hedonic outcomes of novel situations by mentally recombining elements of prior experiences into possible scenarios, an ability which has been termed *affective forecasting* (Wilson and Gilbert 2003). Standard examples for affective forecasting are gustatory predictions, for instance, the prediction that lemon juice with sugar will be tasty, but that a popsicle made of liver will be disgusting. Importantly, individuals capable of affective forecasting are said to be able to make these assumptions without having tasted lemonade or liver popsicles before.

Affective forecasting, with its emphasis on the pre-experience of future feelings, is commonly regarded as a particularly clear example of episodic mental space travel simulation. It has repeatedly been argued to be a uniquely human ability, with animals only being capable of predicting the outcomes of familiar scenarios, if at all (Gilbert and Wilson 2007, Suddendorf 2017). If this hypothesis were correct, animals would have to rely on trial-and-error learning or should be expected to act in accordance with a semantic rule or heuristic, when simulating novel scenarios.

In attempting to falsify this hypothesis, Sauciuc et al. (2016) confronted an orangutan and 10 human participants, who served as control group, with 6 novel juice blends, each of which was made from 2 familiar juices (for instance, cherry and lemon). The subjects could then choose to drink the novel liquid (cherry-lemon) or another familiar juice (for instance, rhubarb). Because prior experiments had determined each participant's exact preferences, Sauciuc and colleagues were able to predict the choice the subjects should rationally make in each one of the 24 unique trials. If the ape, just like the human participants, was able to mentally 'pre-taste' the novel juice blends, his choices should therefore be comparable to those of the human participants.

The results matched this hypothesis. In a rather complex experimental procedure, the authors also controlled for, and excluded, various alternative explanations, for instance, familiarity and novelty biases or preferences for certain colors (ibid. 1090). Importantly, the ape's choices were also inconsistent with a rule-based account, according to which he would always choose the vessel containing his most favorite juice (cherry) or always avoid his least favorite one (lemon). In contrast, the ape chose, for example, rhubarb over cherry-lemon or cherry-vinegar, but not over cherry-rhubarb.

Taken together, the results of the study led the authors to conclude that the orangutan's choices were indeed based on his predictions of the hedonic outcomes of the liquids and that he, consequently, was capable of affective forecasting, challenging the hypothesis that affective forecasting was an exclusively human ability (even though the authors acknowledge the severe limitations of a study with only one non-human participant[98]).

5.1.3.5 How Far-sighted is Great Ape Foresight?

With regard to possible capacities and limitations in apes' future thinking abilities, it would be interesting to know whether great apes are capable of dissociating two future needs from one another and if they take different delays into account. To my knowledge, no such studies, which would be comparable

98 The problem of the very limited number of participants is not immanent to Sauciuc et al. (2016), but affects many, or even most, empirical studies on animal planning capabilities, as should have become evident by now. The problematic implications arising from these circumstances can be better assessed and evaluated by somone other than me, so I will not go further into this complex topic.

to those with jays by Cheke and Clayton (2012) or Thom and Clayton (2014), have been published, however.

In the following paragraphs, I would like to present a study that has been argued to provide "another piece of evidence to the idea that great apes plan for the future" (Bräuer and Call 2015, 262), even though, as the authors themselves acknowledge, it does not meet the requirements of the spoon test. First and foremost, current and future motivational states of the participants do not qualitatively differ and the two instances of the preparatory behavior and the future event are not spatially and temporally separated (for details of the spoon test paradigm, see chapter 5.1.1). It could be questioned, therefore, whether the relevant study *de facto* investigates planning abilities at all. This concern will be readdressed below.

Juliane Bräuer and Josep Call (2015) aimed to inquire whether apes would produce more tools for future use, when it would be rational for them to do so, for instance, in conditions, in which several tools were needed to obtain the reward from an apparatus and in conditions, in which the apparatus was only accessible for a limited amount of time.

One reason for this study was to replicate prior findings and observations on tool-manufacturing of a chimpanzee under controlled experimental conditions. For one thing, Osvath and Karvonen (2012) had observed that Santino – the stone-throwing chimpanzee, living in a Swedish zoo – prepared a stack of projectiles and hid them at several locations, which might generously be interpreted as to indicate his ability to plan several steps ahead into the future. In contrast, the apes in Osvath and Osvath's (2008) study seemingly understood that the function of their tool did not maximize with quantity in that it was not better to posses more than one straw from which fruit soup could be sucked up. Taken together, this raises the possibility that chimpanzees might not only able to plan for a specific future situation, but might even think ahead several steps in advance.

In order to replicate Osvath and Karvonen's (2012) finding and to test whether Santino might indeed have planned several steps in advance, when manufacturing and hiding multiple projectiles, Bräuer and Call (2015) were interested in whether chimpanzees, bonobos and orangutans would manufacture more than one tool in advance, when it would become useful for them to do so.

The procedure of the experiment was as follows. The apes learned, first, that they could make a stick out of a wooden board, which could then be used to reach for a grape from a transparent tube, but also, second, that the stick could not be retrieved from the apparatus afterwards. Thus, each manufactured tool could only be used once. The 25 apes (14 chimpanzees, 4 bonobos and

7 orangutans) were then tested on 3 different conditions, that differed with regard to the number of tools that would be needed. In the all-grape condition, all 8 tubes were baited with grapes so that the apes needed 8 sticks to retrieve them all, whereas only one tube was baited in the second condition, requiring one tool. In the third condition, no tool was needed to obtain 8 grapes in total, because the grapes were easily accessible. The experiment consisted of 45 trials, 15 for each condition, which were presented in randomized order (for details, see Bräuer and Call 2015, 258).

On the first stage of each trial, the apes entered the room with the mounted, but locked apparatus and were allowed to manufacture sticks for 8 minutes. As the apparatus was transparent, the animals had visible access to the tubes and the grapes and were thus able, at least in principle, to know if and how many tools would be needed. Afterwards, the apes were allowed to insert the sticks they had previously made. This phase lasted for either 120 or 50 seconds.

If the apes were not only capable of manufacturing tools in advance, but could also anticipate how many tools would be required, they would be expected to produce 8 tools in the all-grape, one in the one-grape and none in the empty-apparatus condition. As, however, the wooden boards were not removed prior to the reward phase, the apes could produce additional sticks during this last experimental phase, which they did. However, the authors hypothesized that when the reward phase is decreased to only 50 seconds, the apes would prepare more tools in advance, especially in the all-grape condition, where they would require 8 sticks in total.

Yet, the apes were found to produce very few tools during the tool preparation phase over the course of the experiment, with the majority of tools being manufactured during the reward phase. On the all-grape condition, the apes manufactured slightly more tools in advance than on the other conditions, but still less than 2 tools on average. Even more surprisingly, the apes did not change their tool-preparing routine, when the period in which the apparatus would be accessible dramatically decreased from 120 to 50 seconds. Thus, the shorter delay did not increase the number of tools prepared in advance.

These results are notoriously difficult to interpret, even more so, because a production of tools in advance did not automatically lead to greater success; in fact, those apes that had produced tools in advance did not retrieve significantly more grapes from the tubes in the all-grape condition than those who had not (ibid. 261). It remains a possibility, therefore, that even if the apes were aware of the future reward phase being shorter than in comparison to previous encounters with the apparatus, this might not have affected their performance.

Maybe most puzzling about these results is that the apes proved to be so insensitive to the specific circumstances of the different conditions. Even though they had visible access to the apparatus in all trials, they did not even come close to manufacturing the required number of tools in the all-grape condition. Even more astonishing, they prepared more tools when only one tool was required. Redshaw and Bulley suggest, therefore, that the apes "may have been producing [the tools] on a rough rule (e.g., 'more visible grapes = make more tools') rather than the precise requirements of the task (i.e., 'make one tool per visible grape')" (Redshaw and Bulley 2018, 14[99]). Similarly, the authors of the study suggest:

> Another more speculative possibility is that failing to match the number of pre-prepared tools to the potential rewards that one could obtain represents a limitation in apes' planning abilities. In other words, apes might know that they need to prepare a tool in advance but anticipating how many tools are needed might be too taxing, particularly for species that do not spontaneously enumerate objects. Future studies are needed to investigate this possibility. (Bräuer and Call 2015, 261)

Indeed, it would be particularly interesting to know whether this potential limitation in the planning abilities of apes should be ascribed to the apes' difficulties in number representation or whether it points towards an inability to think several steps ahead into the future. This latter ability has been argued to be particularly relevant for strategic planning. Due to the experimental setting, enabling the animals to continue to retrieve food rewards even if they had not prepared wooden sticks in advance, the advantage of anticipating how many tools would be required may have been too insignificant as to motivate the animals to manufacture tools during the first phase of the experiment in the first place (ibid. 262).

More crucially, however, is that even if subsequent studies were accordingly modified, as Bräuer and Call propose, it should be seriously doubted whether positive findings would indicate any planning abilities on behalf of the animal. Because it does not meet the requirements of the spoon test, the current experiment is unable to make any claims as to the animals' anticipatory planning capabilities. Nor, however, is it in a position to make any (non-)affirmative statements regarding the animals' strategic planning abilities, because it does not test for their ability to mentally try out alternative event sequences

99 The page numbers of the cited article refer to a revisioned and edited authors' manuscript of the paper, accessed via ResearchGate on September 12, 2017.

in advance of taking action. As has been argued above, temporal updating enables an individual to simulate a short sequence of events in advance of action in the absence of the ability to *reason* about their order. This latter ability, however, that is, temporal-causal reasoning, has been argued to be crucial for strategic planning.

To conclude: that the great apes in Bräuer and Call's study only scarcely prepared tools for later use, does not make any negative claims regarding potential limitations regarding their planning abilities. Neither, however, would a more positive finding allow the conclusion that the apes indeed are capable of planning several steps ahead in time.

5.2 Behavioral Investigations of Strategic Planning in Animals

5.2.1 *Of Towers and Tools*

As a quick reminder from what has been laid out in more detail in chapter 4.1, it is not being disputed that most of the behaviors and actions animals and humans alike are performing in the present, function to serve a certain future need. And there are also relatively many animals who can behave in a future-oriented manner by executing several actions in a specific order. This, however, does not necessarily represent an underlying capability of strategic planning. As the example of the digger wasp should keep on reminding us, even inflexible, fixed action patterns can look impressively complex at first glance. Even more significant is that even if those inflexible, future-oriented behaviors are ruled out, an empirical investigation of strategic planning requires still more. Strategic planning is not just the ability to execute a sequence of actions, even if these actions are performed in an intentional, goal-directed rather than a merely goal-oriented manner (Saidel 2009).

Take one of the most famous examples in the comparative cognition literature, namely Wolfgang Köhler's chimpanzee Sultan, who stacked several boxes onto each other to obtain a banana, that had been attached out of his reach under the ceiling. Did Sultan engage in strategic planning? By help of the example alone, this question cannot be answered. In order to provide strong evidence that an individual is capable of strategic planning, he must have reasoned about the causal significance of the temporal order of events, that is, he must be capable of temporal-causal reasoning (McCormack and Hoerl 2011; see chapter 4.2.3 for details). Again, this means that the individual is not necessarily required to mentally try out the *complete* strategy beforehand or think about *every* possible alternative course of action that there is (see also chapter 4.4.4.4), but rather must be shown to mentally represent action alternatives and, thus, to have the ability to mentally fill out one time slot with different actions or events and their different outcomes.

One consequence from this is that behavioral investigations of strategic planning must necessarily require the animal to reason about the order of events. In human participants, strategic planning has been investigated by help of, for instance, the Tower of Hanoi or Tower of London task. In animals, typical examples that have been argued to provide evidence for planning capabilities are sequential tool-use experiments or maze tasks. Below, some examples for sequential tool-use and maze tasks will be laid out in more detail. Afterwards, the studies' explanatory power regarding strategic planning abilities and possible limitations in animal cognitive capabilities will be discussed. In forestalling the conclusions I will make below, it will be argued that neither of the sequential tool-use tasks nor the maze tasks provide enough reason for the claim that animals are capable of strategic planning.

5.2.2 *Empirical Studies*

5.2.2.1 Sequential Tool-use in Corvids and Great Apes

Among the many studies investigating sequential tool-use in animals, two will be discussed in the following paragraphs. The first one of these was conducted by Joanna Wimpenny and colleagues in 2009 and represents the first demonstration of a spontaneous three-step sequential tool-use in a non-human animal, namely, a New Caledonian crow. Three years later, Gema Martin-Ordas and colleagues published their results on a very similar task with great apes. The two studies have the major advantage that they are directly comparable to one another, therefore.

At first glance, sequential tool-use seems to provide a clear-cut case for strategic planning abilities. That a crow, after having closely inspected a transparent apparatus with a trapped food reward, uses an available short stick, for instance, to retrieve a stone, which can be used to retrieve another object, and so on, until the bird finally is able to retrieve a piece of meat with the help of a large wooden stick, is undoubtedly a very impressive sight. One crow has even been shown to correctly execute eight steps in a sequential tool-use task, using 5 tools in total (Inside the Animal Mind: The Problem Solvers 2014[100]). To succeed with tasks like these, it seems that the animal must necessarily plan ahead several steps and bring the required action steps into a correct order.

However, sequential tool-use behavior is often alternatively explainable by various other simpler cognitive processes. If the animals previously acquired

100 The 8-step sequential tool-use task was part of a BBC Two documentary called *Inside the Animal Mind* (episode 2) from 2014.

all the necessary behavioral steps that are needed to solve a task, for example, then they may simply 'chain' these separate behaviors into a sequence; especially, if the animals do not have to decide between multiple equally accessible objects, that is, in cases with only limited action alternatives. Wimpenny and colleagues consequently come to the conclusion that two famous studies, examining sequential tool-use in rooks (Bird and Emery 2009) and New Caledonian crows (Taylor et al. 2007) did not suffice to provide clear evidence for (strategic) planning capabilities in these animals (Wimpenny et al. 2009, 1f).

For this reason, **Wimpenny et al. (2009)** included several controls in their experimental set-up. First of all, the New Caledonian crows (*Corvus moneduloides*) were presented with multiple tools in order to test for their tool-selectivity. If the birds were able to plan ahead, they should only retrieve those tools in each task that they would actually *need* to rule out that tools had been associated with a general utility in previous tasks. Second, and in contrast to previous studies, Wimpenny et al.'s birds were sometimes required to use 3 rather than just 2 tools in sequence to obtain the food reward. The reason for the inclusion of 3-step tasks was that the more remote the goal state is from the initial action, the less probable a correct performance at random becomes. Furthermore, Wimpenny et al. (2009) included several control conditions, where either no food was present in the apparatus (to test whether the birds would still retrieve and attempt to use the tools), the tools were replaced with non-functional items (in which case the birds should not retrieve them) or the food reward was accessible without sequential tool-use (where the birds should show no interest in the trapped tools).

To investigate the role of pre-training, the 6 tested New Caledonian crows were furthermore separated into 2 groups, the experienced and the inexperienced. The 3 birds of the first group were allowed to gain experience in using their beaks to retrieve tools from tubes, which could then be used to get a food reward prior to the experimental trials, whereas the other 3 subjects were merely allowed to inspect the empty tool-tubes and use a freely available tool to obtain a food reward from the food-tube. The inexperienced birds, thus, did not learn to retrieve tools from the tool-tubes, in contrast to the experienced individuals. Both groups, however, were allowed to use a tool in order to reach a food reward prior to the task (Wimpenny et al. 2009, 4).

The experiment included 2 different 2-step tasks and one 3-step task, which were not presented according to their increasing complexity but at random. Depending on how deep in the tube the food reward was, 2 or 3 steps were necessary to retrieve it. In the easiest condition (first 2-step task), the birds were required to use the freely available tabletop tool to retrieve any one of the 4 tools, 3 of which were 10 cm and one 20 cm in length, that were each trapped within tubes. The new tool could then be used to reach the reward in the food-tube. In the more difficult condition (second 2-step task), only the longest of the trapped tools (20 cm) would turn out to be long enough to retrieve the food reward. Finally, in the most difficult experimental condition (3-step task), the longest tool (20 cm) was out-of-reach of the tabletop tool, so the birds were required to use the tabletop tool to retrieve one of the 3 medium-length tools and then use this tool to get the longest one, which then provided access to the food reward. In total, 12 trials, 4 per task, were conducted.

The main results were as follows. None of the 3 inexperienced birds succeeded only once on any of the 12 trials. One bird turned out to be either aversive or inattentive to the tool-tubes, so his performance may not necessarily reflect an inability of planning. Another individual interacted with the tool-tubes on several occasions and even retrieved tools with the available tabletop tool, but failed to further use this tool for food extraction (ibid. 6). This indicates that this bird did have a grasp of how the available object could be used, at least on some occasions, but was seemingly unable to further integrate this piece of information. Even worse: on the only occasion, when an inexperienced bird performed a correct 2-step sequence, using a previously extracted tool to probe into the food tube, he did so in one of the control conditions, when the tube contained no food reward at all. In a nutshell, there was no reason to assume that the inexperienced crows had mentally represented the required sequence of actions beforehand.

In contrast to the surprisingly poor performance of the inexperienced birds, all crows, who had previously learned to use the tabletop tool to extract another tool, showed sequential tool-use in their very first trial and consistently afterwards. (One bird, however, did not consistently use the tabletop tool, but brought his own tool on some occasions. As his tool was longer than the available tabletop tool, he invented a clever way to circumvent the need to use multiple tools, thus.)

In spite of the overall good performance of the three crows, the authors also observed extreme individual differences within the group of the experienced birds. Whereas 2 (and another bird from the inexperienced group, who was given additional trials later) succeeded fairly well on the 2-step tasks (ibid. table 2 for details), but did not learn to solve the 3-step task over the whole course of the experiment[101], one bird's performance excelled all others. The crow Betty was not merely the only one, who solved the 3-step task at all, but

101 A second experiment found that only after many trials, the birds eventually learned to solve the 3-step task. However, individual performances were still found to be varying to a very high degree, with some birds being much more successful than others.

even succeeded in 100 percent of all trials in the experiment. This raises the question, whether Betty indeed planned her next moves ahead in time and whether she was capable of strategic planning.

The results of Wimpenny et al. (2009) are remarkable in many ways. On the one hand, Betty's consistently successful performance and especially her spontaneous demonstration of a 3-step sequential tool-use raise the possibility that she indeed planned her actions in advance. On the other hand, it is stunning that even after so many exposures to the task, other birds still continued to make many mistakes on the 3-step task, which seems to indicate a lack of insight into the specific requirements of the task and might, thus, reflect an inability to plan several steps in advance. In a similar fashion, the authors emphasized that the crows made certain errors that are not compatible with a "fully rational planning" explanation (ibid. 12).

Before I make the case that Betty should still not be ascribed with strategic planning capabilities, the results of **Gema Martin-Ordas et al. (2012)** will be summarized, who used a very similar experimental set-up to investigate sequential tool-use in great apes, allowing a direct comparison to the performances of the New Caledonian crows.

The study tested 8 chimpanzees (*Pan troglodytes*), 3 bonobos (*Pan paniscus*) and 4 orangutans (*Pongo abelii*) on a task, whose set-up was very similar to Wimpenny et al.'s (2009) first experiment, which has been discussed above. A major difference was, however, that none of the apes received any degree of pre-training or familiarization with the relevant objects prior to the experimental trials. Despite this additional challenge, the apes succeeded on retrieving the food reward significantly above chance in their first trial in both 2-step conditions and quickly learned to solve the 3-step task, in stark contrast to the crows (see Martin-Ordas et al. 2012, table 4 and below for details).

As had been the case with the New Caledonian crows, however, there were great individual differences. 4 apes solved the 3-step task in their first trial and also succeeded in 100 percent of trials in all conditions, whereas 3 individuals performed extremely poorly in any sequential tool-use task, regardless of whether they had to use 2 or 3 tools in the correct order. The remaining 8 individuals eventually solved the 3-step task in their second, third, fourth and fifth trial, respectively.

It is important to note that once they had learned to solve the 3-step task, the apes consistently used this strategy over the course of the experiment – with one interesting exception: the chimpanzee Sandra solved the 3-step task very quickly, that is, in her second trial. Yet, she did not maintain the successful strategy in subsequent trials and in total succeeded in only 33.33 percent of the 3-step condition. Her performance may raise doubts, therefore, as to whether she actually engaged in a goal-directed reasoning process at all. The other 7 individuals successfully retrieved the food reward on subsequent 3-step tasks, once they had managed to solve such a task for the first time.

In sum, Martin-Ordas et al.'s results suggest that the apes took the specific requirements of the tasks into account very well and acted accordingly (ibid. 8). Their performance is even more remarkable in light of the fact that they had not been trained in any component of the sequence prior to the experiment. They solved both 2-step conditions significantly above chance in their first trials, whereas the (experienced) crows had solved these only in their fourth and fifth trials, respectively (ibid. 13).

In a second experiment, Martin-Ordas and colleagues found that the apes could even use up to 5 tools in a sequence. One chimpanzee named Fifi turned out to be a particularly good sequential tool-user as she was the only individual who succeeded on 100 percent of the trials in both experiments. Fifi was not the only one who spontaneously solved the 5-step sequential tool-use task, but she was the only one who consistently succeeded in all subsequent trials.

In conclusion, two individual performances stand out among the others in both studies that have been sketched above. Among the 6 New Caledonian crows, Betty was the only one who solved the 3-step task from her first trial on. Chimpanzee Fifi even did so without having been pre-trained on the task before. In the following paragrahs, these two performances will be examined more closely with regard to their explanatory power concerning the animals' underlying planning abilities. As impressive as the performances of Betty and Fifi are, I will argue that they do not provide evidence for underlying strategic planning abilities.

To clarify: in chapter 4.2.3, it was argued that animals might very well be able to mentally simulate a short sequence without simultaneously being able to reason about the *order* of these events (see also McCormack and Hoerl 2011, Hesslow 2002). This latter ability, that is, temporal-causal reasoning, however, has been argued to be a necessary requirement for strategic planning in chapters 4.4.2 and 4.4.3. The experimental set-up in the two studies by Wimpenny et al. (2009) and Martin-Ordas et al. (2012) does not require the individuals to mentally put the involved actions into a temporal order. The results do not suffice to argue that animals are capable of strategic planning, consequently.

Admittedly, both the crow and the ape were required to decide between two equally available tools at the beginning of the action sequence. Their correct choice, thus, can at least be interpreted to represent the knowledge that one tool would enable them to reach another. It should be doubted, however, that this choice is enough to indicate any planning abilities on behalf of the individual. If one item is functional in that it allows the retrieval of another object, but the other one is not, it should be quite rational for the animal to choose the functional over the non-functional one.

If the simulation of one course of action leads to a dead end, then it is plausible that the animal will not put this mental representation into action. If, in contrast, a simulated course of action will have a positive effect, that is, the retrieval of another object, then it should be rational to assume that the animal will put this mental representation into action. If it is acknowledged that the simulation of a short sequence of events can occur via a simple simulation mechanism, which is compatible with a temporal updating explanation, then this is simply not enough to assume temporal-causal reasoning and, accordingly, strategic planning.

It is important to note that in simulating this short action sequence, the animal might very well have done so via mental space travel, which is compatible with a temporal updating explanation. It is plausible to assume, for instance, that Betty simulated herself probing the tool tube with the longer tabletop tool and retrieving the longer tool, prior to taking action. (Maybe Betty even first simulated herself unsuccessfully probing the tool tube with the shorter tabletop tool.) I acknowledge, thus, that the animals in the above experiments might indeed have simulated a sequence of events via mental space travel and maybe they even first simulated one course of action and then another.

Crucially, however, this does not automatically indicate a grasp of temporal-causal relationships, because an individual capable of successively simulating two sequences of events does not necessarily understand that these two simulations represent two mutually exclusive options of how to act. Put differently, I argue that Betty and Fifi were indeed required to correctly choose one (functional) object over another, or one way to act over an alternative option, but they were not required to reason about the temporal order of two events, that is, to "mentally try out the consequences of events occurring at different temporal locations in a sequence" (McCormack and Hoerl 2011, 145).

5.2.2.2 Mind the Trap: Navigation through Mazes

There are numerous experiments researching how animals navigate through mazes. Many of these tasks argue that to provide evidence for animal planning abilities on the assumption that in order to succeed, the animals must mentally represent their steps in a top-down manner before executing their first move. As I will argue below, however, it can be doubted that these experiments indeed allow inferences to any underlying strategic planning abilities on behalf of the participants. In doing so, I will focus on one recent example for such a maze task, that carefully differentiated between several task conditions and controlled for several possible explanations, that is Michael Beran et al. (2015). In comparison to another maze task study by Emma Tecwyn et al. (2013), this study has several clear advantages, as I will attempt to explain in the following paragraphs.

The original set-up of the so-called paddle box[102] in **Tecwyn et al.'s (2013)** first experiment required the participating bonobos (*Pan paniscus*) and orangutans (*Pongo pygmaeus*) to adjust the position of the paddles *prior* to moving the food paddle, holding the food reward. The demands of this task were too high, however: in the vast majority of the cases, the apes exclusively rotated the food paddle but no other paddles, which led to the food reward becoming inaccessible in the majority of cases.

To rule out that the apes' poor performance was due to an inability to inhibit the tendency to immediately rotate the food paddle or to a lack of understanding how diagonally positioned paddles affected the path of the reward, the second experiment included some changes. The main difference to the first experiment was that all paddles were now set up in a flat rather than a diagonal position. This ensured that all trials on this experiment could be solved by always only rotating the respective paddle containing the food reward. The subjects could thus proceed in a step-by-step manner, with the food reward only falling from one level onto the next lower one. Most apes performed well in this modified experiment.

The modified set-up of Tecwyn et al.'s second experiment has several problematic implications, however, that make it impossible to answer whether the apes indeed thought through all the necessary steps prior to taking action. Especially in conditions with multiple solutions, the individuals could just proceed by making one decision at a time, that is, by solving the task in a purely opportunistic manner rather than by being required to think several

102 The paddle-box apparatus consisted of an opaque box, containing 8 rotatable paddles, 3 at the top level, 2 in the middle, and 3 at the bottom level. The current position of each paddle (flat, diagonal left or diagonal right) could be changed by the subject by rotating a wooden handle, extending from the paddle itself. At the bottom of the apparatus, there were 4 possible goal locations, which could be opened or locked. The specific design of the apparatus made it possible to adjust the complexity of the task. In the easiest condition, a food reward was put on 1 of the 3 paddles at the bottom level with the animal having to determine whether to rotate the paddle to the left or the right, depending on which of the 4 goal locations was open (1-step task). If the food reward, in contrast, was located on a paddle at the highest level of the paddle-box, the animal sometimes needed to rotate up to 3 paddles to obtain the reward (3-step task). On some conditions, there was only one solution of the task, whereas in other conditions, there were several ways to successfully retrieve the reward.

steps ahead at all. In tasks with only one possible solution, where the participants were, at least in theory, required to think several steps ahead, the subjects could succeed on 100 percent of the trials by simply moving the paddles towards the location of the goal – the Euclidean direction – which makes a top-down strategic planning approach only *one* among several options to explain the animals' success on these tasks.[103]

In sum, Tecwyn et al.'s (2013) results are not telling with regard to strategic planning abilities. Because the few relevant tasks, that seemed to actually require the apes to think ahead, could alternatively be solved by always rotating the paddles towards the goal location, the possibility that the apes applied a procedural rule in these tasks cannot be excluded. To make progress on the question whether great apes are capable of thinking ahead and if so, to what extent, it would be particularly enlightening to focus on the kind of errors the apes may make on these and similar kinds of tasks and to analyze whether there may be patterns that could point towards certain limitations.

This is exactly what **Michael Beran and colleagues (2015)** were interested in finding out. Their study has the additional major advantage of testing not only several primate species but also including children between 2 and 5 years of age, which makes it possible to directly compare their performances with those of the primates. In detail, the authors tested 4 chimpanzees (*Pan troglodytes*), 7 rhesus macaques (*Macaca mulatta*), 8 capuchin monkeys (*Cebus apella*) and 26 human children (*homo sapiens*) on a computerized maze task, that shared certain features with the paddle paradigm used by Tecwyn et al. (2013), but also contained important differences. The mazes consisted of 4 levels, with the cursor starting somewhere on the top level on each trial and a goal location at the bottom. With a joystick, the subjects could move the cursor either to the right or left. When the cursor reached a gap, it 'fell' down until it hit solid ground again or dropped onto the bottom of the maze without touching the goal, which resulted in the game being lost. Thus, the subjects were required to avoid those gaps, that would result in the cursor hitting the maze bottom instead of the goal, but seek those gaps, that would enable them to safely reach the goal location. In contrast to the paddle box, several mazes also contained traps, from which the subjects were unable to free the cursor for the rest of the

103 Even though the authors emphasize that at least the orangutans did not *always* move the paddles in the Euclidean direction (Tecwyn et al. 2013, 182), it would be compatible with the study's results to assume that they applied an alternative heuristic to do so *whenever they did not know the correct next step*, that is, in multiple step-tasks, where there was only one correct solution available. This would also be in line with the results from experiment 1, where the orangutans showed a consistent preference for rotating the paddles towards the goal.

game. A successful completion of each trial resulted in a food reward, while an unsuccessful attempt was followed by a time-out.

Similar to Tecwyn et al. (2013), the authors classified all 100 mazes, that were used on this study, from easiest to most difficult. One way to classify the mazes would be to count the number of critical choices the mazes contained, that is, choices, that "led to a guaranteed (and irreversible) error, such as missing the goal when the cursor finally reached the bottom of the maze, or falling into a trap" (Beran et al. 2015, 8). However, the authors did not consider the number of critical choices to be decisive for the complexity of the task, but whether or not it required a reversal of cursor direction or a movement against the Euclidean direction, that is, away from the goal. A particular maze, for instance, could make 3 critical choices necessary, but be solved by simply always moving the cursor in the same direction, namely towards the goal (which would be equivalent to Tecwyn et al.'s 3-steps-1-solution paddle box tasks). Such a maze would still be classified as *easiest* on Beran et al.'s categorization, despite its high number of critical choices.

Thus, rather than separating the mazes with regard to their number of critical choices, the mazes were separated into the following 4 categories: *direct trials* could be completed by simply always moving the cursor in the direction of the goal, whereas in *Away trials*, the cursor sometimes needed to be moved away from the goal. In *Reversal trials*, the subjects needed to reverse the direction of the cursor sometimes. A reversal of direction was not necessarily equivalent to a movement away from the goal, as can be seen in Figure 2 of Beran et al. 2015, but if it did, the relevant maze was separated into the fourth category, which contained *Reversal + Away trials*.

The authors hypothesized that Reversal trials would prove to be especially difficult for those individuals struggling with motor control or inhibition, whereas Away trials were expected to be particularly challenging for individuals unable to plan multiple steps ahead, because they required the anticipation that a certain move towards the goal location would lead to an irreversible error a few steps later (ibid. 5). Based on previous findings, the authors expected the monkeys to perform worse than the chimpanzees on all trials that required a reversal of direction and/or a movement in the opposite direction of the goal location. Overall, the chimpanzees were expected to reach about the same performance level as children between 2 and 5 years of age, even though the older children in the study should perform better than the younger ones. To test for any learning effects, each maze was presented twice to each individual, which resulted in a total of 200 trials.

The results matched the above hypotheses. Each of the 4 chimpanzees correctly solved more than 80 percent of the mazes overall. Impressively, there were no statistical differences between the 4 type trials, that is, the chimpanzees performed more or less equally well on the easiest and the hardest maze tasks, even though, as the authors point out, the size of the sample was very small in the case of the chimpanzees and, importantly, all individuals had had prior experience with similar maze tasks. A closer look on individual results also reveals that performances generally decreased in the more complex trials, especially in case of one individual, Lana (see Beran et al. 2015, Figure 4). Still, "overall, the chimpanzees seemed to show clear evidence that they were investigating routes through the maze, because they anticipated the need to move away from the goal sometimes, and they also showed the motoric inhibition needed to execute the reversals" (ibid. 14).

Both monkey species performed quite well in Direct trials, but their performances significantly decreased in the other 3 trial types. A movement away from the goal location was found to be particularly challenging, even more so, if the animals were additionally required to reverse directions during the task. Consequently, Reversal + Away trials were performed at the lowest level; especially by the capuchin monkeys. The authors concluded:

> This indicates that monkeys showed both inhibitory difficulties and difficulty anticipating trouble spots that require moving away from the goal. A comparison of performance on the first and second presentation of each maze indicated that performance did not improve with experience with a specific maze, and so this may be a real limitation in monkey planning abilities. (Beran et al. 2015, 14)

Beran and colleagues point out, however, that more research will be needed to strengthen this hypothesis. First and foremost, it would be interesting to know whether eye-tracking would support the claim that chimpanzees actually think ahead in a top-down manner, whereas the monkeys do not or to a much lesser extent. If the chimpanzees think ahead several steps in advance, their gazes should be expected to 'wander' through the maze prior to making a decision at a critical point. The monkeys would not be expected to do so. Furthermore, monkeys and apes should be directly compared to one another on other planning paradigms besides maze tasks, as the authors repeatedly emphasize.

The 26 children, who participated in the study, were separated into 2 groups, with 13 children being between 28 and 48 months of age (i.e., between 2.5 to 4 years) and 13 between 54 and 66 months of age (i.e., between 4.5 to 5.5 years). Many children correctly solved 80 percent or more of the trials in total, just like the chimpanzees did, but older children performed better than younger ones. Furthermore, and in contrast to the chimpanzees, performances significantly decreased with the tasks getting more complicated.

If the respective performances of chimpanzees and children in the 4 separate trial-types are compared to one another, it becomes evident that chimpanzees and children performed about equally well on Direct trials. In the remaining 3 trial-types, however, the chimpanzees generally performed better than the children. Apes and older – but not younger – children were about equally successful on Away trials. In trials that required a movement of the cursor in the reversed direction, however, the chimpanzees performed better than both groups of children. Especially young children, just like the two monkey species, had particular difficulties in those tasks that required them to move the cursor in the opposite direction and away from the goal location, that is, in Reversal + Away trials, which resulted in a considerable decrease of successful performances (Beran et al. 2015, Figure 5). These results might explain the authors' conclusion that the performances of the children in some ways paralleled those of the monkeys, but in other ways came closer to those of the chimpanzees (ibid. 13).[104]

The 4 chimpanzees in the study and the children of 4.5 years of age or older had no difficulties in solving the different kinds of mazes, even when they were required to move the cursor in the opposite direction or away from the goal – even though the chimpanzees were the only group showing no significant differences among the trial-types. It seems safe to conclude, thus, that both chimpanzees and the older children made their decisions on critical choice points in the maze in anticipating potential traps or incorrect routes and successfully avoided these. Rhesus macaques, capuchin monkeys and younger children, in contrast, struggled.

How do these results now relate to the ability of what has been called strategic planning? In our daily use of language, typical examples of strategic planning are often more medium- or long-term, such as career planning. However, the temporal horizon should not be decisive for whether or not an animal is being ascribed with a certain capability or not. Still, it will be argued that successful maze task performances do not provide sufficient evidence for strategic planning abilities.

---

104 One must be cautious not to make too direct comparisons between the nonhuman primates and the children in this study, however, because there were important methodological differences, as Beran et al. (2015) themselves emphasize. The 4 chimpanzees were not only experienced with this kind of task, but had also received extensive training prior to testing, whereas the children were verbally informed about the goal of the task. Furthermore, the chimpanzees, but not the children, were rewarded with food for correct responses, which might have affected motivation. Nonetheless, these results may be illuminating with regard to potential cognitive limitations in these different species.

The Tower of London test has been traditionally used to investigate (strategic) planning abilities and possible deficits in humans. In its original version, which was developed by Tim Shallice and Rosaleen McCarthy, 3 colored beads, resting on 3 sticks of differing heights, need to be brought into a specific target configuration in the minimum number of moves (Shallice 1982). To reach the goal state, the subject needs to be aware that sometimes beads will be blocked by another one resting on top of it. Due to its specific design, the Tower of London cannot be solved via a mere understanding of which moves must be made, therefore, but only in which *order* these moves must be made (McCormack and Hoerl 2011, 131).

According to McCormack and Hoerl, "[p]lanning in tasks such as the Tower of London requires something like the temporal equivalent of arranging furniture in a room: it requires mentally arranging a sequence so that moves are placed in specific temporal locations relative to each other" (McCormack and Hoerl 2011, 131) and note that this goes beyond a mere representation of events *in* a certain order. Thus, the Tower of London is a good empirical measurement to investigate temporal-causal reasoning or, more generally, event-independent thinking.

Temporal-causal reasoning has been argued to be a crucial element of strategic planning. A plan is not equivalent to a to-do list, I have argued, but involves the ability to "reason about the causal significance of the temporal order of events" (McCormack and Hoerl 2007, 168), that is, an understanding of how the order, in which certain things are being done, will affect the outcome.

To successfully find one's way through the mazes discussed above, however, does not presume temporal-causal reasoning abilities, because the events in the maze cannot be brought into a temporal order, to begin with. They require something like the ability to think several steps ahead in space, but this is not equivalent to being able to mentally identify and temporally organize alternative courses of action. The relevant decision to make in the above maze tasks is which of two or more possible ways must be taken but not, in which temporal order several actions should be executed. Therefore, Tecwyn et al.'s (2013) and Beran et al.'s (2015) results do not provide evidence for strategic planning abilities in great apes or infants.

Against this background it becomes particularly interesting that chimpanzees and young children successfully manage to find routes through mazes and think ahead insofar that they avoid traps and wrong routes, but have not yet been shown, to my knowledge, to solve the comparably 'easy' – from an adult human person's point of view – branched-tube task. How can this be explained? Below, I will attempt to unify this surprising finding by suggesting a possible explanation, relying on the kind of concepts that have been presented throughout chapter 4 of this thesis.

One possible interpretation for the chimpanzees' and infants' success in Beran et al.'s (2015) maze tasks may be that chimpanzees and infants do very well know the difference between a correct and an incorrect route, because they mentally try them out in advance of taking action. One route exposes itself as successful and for this reason is followed afterwards.

In the branched-tube task, in contrast, this strategy is not sufficient. Successful subjects additionally need to understand that a potentially correct route will only *sometimes* turn out to be correct, because it only represents one *possibility* of how the future is going to unfold, an ability which has been termed the representation of branching futures. Future research may very well provide evidence that chimpanzees are capable of this kind of thinking, but until then, it remains at least a possibility that chimpanzees mentally try out different routes through a maze in the absence of understanding how these different representations are temporally related to one another, namely as possible, mutually exclusive scenarios of what is (or could) be happening next.

At first glance, this explanation may sound rather far-fetched. Compare it, however, to the findings of Sarah Beck et al. (2006), who tested children on the Y-shaped tube task and afterwards tested their abilities to answer counterfactual and future hypothetical questions. The researchers found that despite being unable to take adequate measures for the mouse or ball to emerge from one exit or the other, that is, to plan ahead for both eventualities, the children could satisfactorily answer where the object will (or would) be if it takes (or would have taken) the other way. If they were asked open counterfactual and hypothetical questions, however, the children struggled considerably. As has been laid out in more detail in chapter 4.2.4, the authors suggested that the children engaged in a mental representation of an alternative version of reality, without simultaneously representing the temporal relationship between both representations, that is, the imagined scenario and current reality.

The chimpanzees may similarly be able to mentally try out different kinds of scenarios and opt for the one, that yields the favored result. Importantly, this does not also imply, however, that the individuals experience these scenarios as alternative *future* realities, nor that they are able to reason about the temporal order of events. If temporal-causal reasoning is being acknowledged as an essential element of strategic planning, however, then even those maze tasks, where the subject needs to think multiple steps ahead in order to avoid traps or incorrect routes, are not sufficient to empirically verify strategic planning abilities.

McCormack und Hoerl (2011) propose to test strategic planning in animals along an earlier devised planning task by Kerry McColgan and Teresa McCormack (2008). In this task, children were informed that a doll, Molly, wanted to take a photograph of a kangaroo on her zoo visit. The children were asked to place the camera on the game-board so that Molly would be able to take a photograph of the kangaroo. This task required the children, first, to reason about the significance of putting the camera at a location, that the doll would be visiting *before* reaching the kangaroo, and, second, to define such a location then. McColgan and McCormack found that children younger than 5 were unable to solve this task above chance, even in an extremely simplified setting with only 2 possibilities of where to place the camera (one before, and one after the enclosure of the kangaroo).

McCormack and Hoerl (2011) suggest to design a similar experiment for animals, in which subjects need to reason about where to place a tool so they would be able to pick it up before actually needing it. If the animals would succeed on such a task, just like the 5-year-old children in McColgan and McCormack's study did, this would prove that they are indeed able to plan ahead strategically (McCormack and Hoerl 2011, 145). To my knowledge, this paradigm has not yet been conducted with animals in controlled experimental settings.

In a nutshell, there does not seem to be sufficient reason to suggest that animals can engage in temporal-causal reasoning, and, consequently, strategic planning thus far. In contrast, it is very likely that some animal species, most notably chimpanzees and jays, possess anticipatory planning abilities. Even though the above-mentioned studies only represent a small fraction of the vast empirical evidence, they allow some important conclusions, I believe, because some of them have been generally taken as providing some of the strongest arguments in favor of animal planning capabilities, whereas others point towards potential deficits and limitations. In the following chapter, these conclusions will be amplified in more detail. Particular focus will be laid on the question how these results relate to the degree of prudential unity relations.

5.3 Intermediate Conclusion

I provided evidence for the claim that some animals are indeed capable of anticipatory, but possibly not strategic planning. In simulating future scenarios, animals may furthermore engage in episodic future thinking via mental space travel. There does not seem to be compelling evidence, however, for mental time travel. In contrast, chimpanzees have been shown to be unable to succeed on branched-tube tasks, which raises the possibility that they do not possess temporal-causal reasoning abilities and are, thus, incapable of mental time travel, as defined in chapter 4.3.

The aim of this chapter is to provide a synthesis of these results with regard to the prudential unity relations that determine how much an individual is psychologically related to her future self. Psychological unity has been argued to occur in degrees, but the difference between strong and weak psychological unity seems to be marked by a threshold on McMahan's time-relative interest account. In a nutshell, a strong psychological unity presupposes the ability to mentally transcend the present, which requires, at its most basic level, an ability to form internal reference-typed psychological connections into the non-immediate future. A typical example for such a connection would be the anticipation of a future desire that qualitatively differs from the individual's current motivational state. Such an individual would be able, in turn, to make sacrifices in not giving in to a more immediate, smaller appetite.

The empirical evidence available suggests that some animals can indeed form these kind of psychological connections into the future and have the potential to form strong psychological unities, consequently. This is in stark contrast to the stuck-in-time hypothesis, defended by McMahan and others, arguing that the mental lives of animals are exclusively restricted to the here and now. Yet, I have repeatedly argued against a simple dichotomy between those creatures stuck to the present moment and the mental time travelers. Two beings with the capability to form internal reference-typed connections, for instance, can still be extremely different from one another concerning their temporal perspectives.

In chapter 3.4.2, I raised the question who really should be described as being stuck in time: the animal – or the philosopher? After making the necessary conceptual clarifications and analyzing some relevant experiments, which, either make a particularly strong case in favor of animal planning abilities or point towards possible limitations, I am in a better position to formulate an answer to this question.

On the one hand, the claim that no other animal besides the human can form internal reference-typed connections to her future self can be dismissed. At least some animals can consciously anticipate future needs and desires and plan ahead accordingly, even if they have to accept present sacrifices. Their future-oriented behavior definitely goes beyond mere fixed action patterns or associatively learned behaviors.

On the other hand, there is no reason thus far to assume that animals may be capable of temporal-causal reasoning and, thus, strategic planning or mental time travel. The philosopher, insisting that there may be certain major differences concerning future thinking abilities between animals and (human) persons, may still have a point, therefore. Adult humans typically do not only think *far* ahead, but also quite *broadly* into the future. Points in time are being represented as empty slots, which can hypothetically hold a variety of different events that might occur or might have occurred then, just like a certain location in a room can hypothetically hold various different pieces of furniture (McCormack 2014). The ability to engage in this event-independent kind of thinking enables a subject to ponder future hypotheticals, but also past counterfactuals, and allows subjects to retrospectively evaluate, and maybe regret, past decisions and even to evaluate decisions that are still to be made, on the basis that they might be regretted later (anticipated regret). Event-independent thinking, therefore, is crucial for what McMahan terms a narrative unity.

Of course, it is plausible that temporal-causal reasoning develops gradually. A first step towards temporal-causal reasoning abilities has been argued to be the representation of branching futures. As has been repeatedly laid out, the ability to simulate alternative future scenarios while being aware that these alternative scenarios represent mutually exclusive options, is crucial for weighing alternative courses of action against each other with regard to their consequences.

To conclude:

- At least some corvid species and great apes seem to be capable of anticipatory planning and can, thus, potentially form strong psychological unities in McMahan's sense, because they are able to transcend the temporally and spatially local, in contrast to the predictions of the stuck-in-time hypothesis.
- These animals very likely simulate future scenarios via mental space travel and are thus able to pre-experience future desires and needs, which has been argued to further psychologically connect them to their future selves.
- A subject capable of mental time travel is even stronger bound to her future self, because she can simulate multiple alternative future scenarios and because she can take several temporal perspectives onto one and the same event, which enables her to experience, for instance, anticipated regret.
- A first step towards showing that animals may be genuine mental time travelers, could be to provide positive evidence for their ability to represent branching futures, for instance, with the help of the branched-tube task. The ability to represent branching futures has been argued to constitute a rudimentary form of temporal-causal reasoning.
- If those animals, solving the branched-tube task, also succeeded on strategic planning tasks, in which they necessarily need to reason about the order of future events, this would strengthen the hypothesis that some animals have temporal-causal reasoning abilities as well.

A central problem of the current research on animal future thinking capabilities is that thus far, primates and corvids have been the focus of attention. This makes it particularly difficult, however, to make any claims with regard to when and under which circumstances future thinking phylogenetically evolved. This is also highlighted by Redshaw and Bulley:

> Absence of evidence is not only a specific problem for certain domains of animal future thinking, but also more a general problem when considering the relatively miniscule number of species that have been tested in controlled settings. The performance of great apes and other primates is of particular interest, of course, given the potential for shedding light on the evolution of human-like future-oriented mechanisms by studying closely related species. Nevertheless, by examining patterns of capacities and limits across vastly different taxa, one could potentially reason about the biological and environmental factors responsible for the emergence of future thinking in general (whether that be 'mere' predictive cognition or higher-order capacities). For example, does future thinking tend to arise as a by-product of domain-general cognitive specialization? Or does it tend to emerge in response to critical environmental pressures such as highly uncertain future rewards or threats that precipitate a need for advanced preparation? And what role do overall brain size, neocortex ratio, or other neurological factors play? These questions will remain moot, however, until more studies are carried out with non-primate taxa other than corvids and rodents. (Redshaw and Bulley 2018, 23)

Ideal candidates would be animal species that have already demonstrated high cognitive performances on other relevant cognitive domains, such as problem solving, or whose behavior makes it at least possible that they may be capable of anticipatory planning. Examples are other mammals, such as cetaceans, dogs or elephants, other bird species, for instance, parrots, and also some invertebrates, for example, octopuses. It is to be expected that at least some of these animal species will be found to demonstrate impressive future thinking abilities as well.

CHAPTER 6

Back to the Ethics of Killing Animals

The purpose of this chapter is to come full circle with regard to the opening question of this thesis, that is, whether animals should be ascribed with a (strong) interest in their continued lives and under which circumstances, if any, it can be morally permissible to kill them. The time-relative interest account (TRIA) measures the badness of a certain death according to the (net)[105] amount of good it denies, multiplied by the strength of the prudential unity relations and, thus, comes to the conclusion that all sentient animals are *ceteris paribus* harmed by their deaths. This means that the killing of animals, whether painless or not, is always morally relevant.

The TRIA evaluates a specific incidence of a token-death, but as it emphasizes the relevance of the strength of the psychological relation that persists between an individual at the time of her death and her future self, it also makes certain assumptions about the general harmfulness of type-death for a specific group of animals in comparison to another group. Thus, the harmfulness of death also becomes a matter of the cognitive capabilities of an animal species. The foregoing two chapters, representing the core of this thesis, argued that some animals can potentially form strong psychological unities, because they are indeed able to transcend the spatially and temporally local by planning ahead and making sacrifices in light of future pay-offs. As a result, a time-relative interest account should ascribe these animals with a much stronger interest in continuing to live than McMahan grants them to.[106]

105 While McMahan argues that the *net* amount of future good should be decisive for determining the loss an individual suffers through her death, i.e., the total amount of good minus the total amount of bad this future life would plausibly have contained, other have argued against this determination of future good. The TRIA has been suggested to be modified to the effect that the loss should be put in relation to what would be maximally possible, in the first place, for the individual in question (see, e.g., Višak 2017 and Cochrane 2007, 2012). Such a modified TRIA aims to determine the suffered losses by a person and, say, a dog from their very own perspectives (see chapter 3.2.2 for details). It should be kept in mind, thus, that a time-relative interest account does not necessarily have to be identical with McMahan's.

106 Even though McMahan considers the possibility that some animal species might not be adequately described as being stuck *entirely* in the here and now, he seems to consider this possibility, first and foremost, with regard to "the higher primates" (McMahan 2002, 265). It seems thus safe to assume that McMahan should at least acknowledge that certain

 | DOI:10.30965/9783957437365_007

The fact that an untimely death does a harm to an animal does not, *per se*, make any *specific* normative claims as to whether we are allowed to kill it. A utilitarian, for instance, may argue that a cow's interest in continuing to live is so weak in relation to our interest in ending the cow's life, that her death can be morally justified. Another person may disagree on the basis that the cow's inherent value gave her a right not to be sacrificed for the greater overall utility. Put differently, the claim that animals are indeed harmed through their deaths remains without normative consequences, if it is not being placed within a moral theory. Before the findings of this thesis can be translated into any moral obligations that we might or might not have towards our fellow animals, I must delve deeper into some of the major positions concerning the ethics of killing animals.

Almost needless to say, a proper investigation of our moral obligations towards animals cannot be done without mentioning two names, that is, Peter Singer and Tom Regan. Since its development in the 1970s, modern animal ethics has moved within the dichotomy of a consequentialist perspective and an animal rights account. Many contemporarily influential theories can still be attributed to one or the other school of thinking, but one should be careful not to fall into a simple pro-Singer versus pro-Regan representation of the matter (see also Cochrane 2012, 7). The positions within one of these two strands occasionally differ tremendously from one another, as will become evident below with the help of two rights accounts. Nonetheless, Singer's and Regan's positions remain to be the most influential theories in the field of animal ethics, and they still serve as a reference point to anyone concerned with it. This may justify my focus on these two particularly influential positions. There are, of course, many more, which unfortunately cannot be discussed here for lack of space.

It is the aim of this chapter to make the conceptual differentiations from chapters 3 and 4 fruitful for animal ethics. In fact, I will argue that any plausible ethical theory of killing should include these differentiations to avoid falling into the simple dichotomy between stuck-in-time creatures versus mental time travelers and which has repeatedly been criticized throughout previous chapters.

I will proceed by first giving a short overview of how the two main positions of modern animal ethics have dealt with the permissibility of the killing of animals and also go into detail of some proponents' views. Afterwards, I will present hybrid accounts, defined here as including utilitarian as well as

---

bird species should be ascribed with a much higher time-relative interest in continuing to live as well (see also chapter 6.3.3).

deontological elements. In chapter 6.4, I hope to show why these theories still are too unspecific. Afterwards, I will sketch a suggestion how my conceptual clarifications may help to solve what I believe to be an essential problem within the field of animal ethics.

6.1 Animal Persons: Peter Singer's Preference Utilitarianism

To be a proper object of moral concern, Jeremy Bentham famously argued, the ability to feel pain or pleasure is sufficient. Beyond that, however, Bentham made no secret of viewing the *painless killing* of animals as morally unproblematic, because animals, in his words, "are never the worse for being dead" (Bentham 1789/1970, 282f Fn b). In a similar fashion, Peter Singer, still one of the most important representatives of a utilitarian animal ethics, concludes his considerations on killing animals with the following sentences:

> As long as sentient beings are conscious, they have an interest in satisfying their desires, or in experiencing as much pleasure and as little pain as possible. Sentience suffices to place a being within the sphere of equal consideration of interests, but it does not mean that the being has a personal interest in continuing to live. (Singer 2011, 119)

This quote might come as a surprise for some who are familiar with Singer's position. And indeed, Peter Singer has developed a rather ambiguous position on the killing of animals, as will become evident in the following.

Singer argues for a *preference utilitarianism*, which regards an action as morally good, if it satisfies as many preferences or (preference) interests – that is, "needs, wants and desires" (ibid. 12) – as possible in consideration of all the parties involved.[107] Presupposed that the arbitrary favoritism of a certain

107 Recently, Singer admitted to lean towards an impersonal hedonistic utilitarianism nowadays, having "bec[o]me more doubtful about preference utilitarianism" (Singer 2015, 234). An action is good, from the point of view of a hedonistic utilitarian, if it maximizes the total utility in the world, regardless of whether this is achieved by maximizing the pleasure of an already existing being or whether by bringing a being into existence, that would otherwise not have existed (Singer 2015, 234). This has also been termed the total view of utilitarianism, in contrast to the prior existence view (Singer 2011). Thus, even though Singer himself does not wholeheartedly defend preference utilitarianism anymore, it continues to represent a major position in the field of the ethics of killing animals and has inspired many other authors. Moreover, Singer has not yet laid out the details of his newer position. For these reasons, the following considerations will focus on Singer's preference utilitarian position, as laid out it in his main work *Practical Ethics*. I will rely on the third edition from 2011, which has been extensively revised by the author.

group of beings is to be condemned[108], then comparable interests, such as the avoidance of pain, must be considered equally. The interest in continuing to live, however, is neither held by all sentient animals, nor is it equally strong among those who have it. In order to have an interest in continuing to life, Singer argues, a being must be self-conscious, defined here along the lines of John Locke's criterion for being a person, that is, as the ability to conceive of oneself as a distinct entity with a past and a future (see also Locke 1694, II.xxvii.9[109]). In accordance with Locke, beings with this kind of self-consciousness are called *persons*, and those, who are "merely conscious" (Singer 2011, 120), are called *non-persons*.

A person's self-consciousness, therefore, provides a direct reason against killing her (ibid. 119). The reason for this lies in the following considerations. To ascribe a (preference-) interest in continuing to live to someone only makes sense, according to Singer, if the relevant being has the ability to form future-directed desires. In order to have future-directed desires, however, the being must be conscious of having a personal future in the first place. A being unaware of tomorrow – *stuck in time*, as it has been called – cannot have any desires or wants that refer to tomorrow. In this regard, Singer's position bears

108 Singer famously argues against *speciesism*, that is, the arbitrary favoritism of all or most members of a certain species, e.g., human beings, over all or most other individuals of another species, e.g., all other nonhuman animals, similarly to racism or sexism. Importantly, Singer does *not* claim that there can *never* be a morally relevant difference between two individuals of two different species. This morally relevant difference, however, cannot be justified with the fact that they do not belong to the same species. Rather, Singer has defended the position that the moral status of each and every individual depends on her particular, individual characteristics and is always to be regarded independently from her group membership. This view has become known as *moral individualism*, based on James Rachels' (1991) definition. Some authors, e.g. Mary Midgley (1983), oppose moral individualism on the basis that the relationship between humans and animals is much more complex than moral individualism grants it to be and that there are multiple morally relevant criteria. Recently, Donaldson and Kymlicka (2013) argued against moral individualism on the basis that we do in fact not have equally strong moral obligations towards certain groups of animals, e.g., wild-living animals versus pets, and, thus, that certain differences between these groups are morally relevant. Relational theories certainly represent an interesting avenue in the field of animal ethics, even though they have problematic implications on their own (for a critical review of Donaldson and Kymlicka, see Cochrane 2013). I do not wish to give a full overview of existing normative theories and positions here, however, but rather carve out certain problematic implications of two major strands, i.e., utilitarianism and rights theories. For lack of space, I will restrict myself to non-relational theories, therefore, i.e., theories arguing in the tradition of moral individualism.

109 The relevant passage can be found in paragraph 9 of chapter 27 of the second book of John Locke's *Essay Concerning Human Understanding*. I refer to a reprint of the relevant chapter in Perry 2008 (33–52).

close resemblance to McMahan's position, as becomes even clearer in the following paragraph:

> For preference utilitarians, taking the life of a person will normally be worse[110] than taking the life of some other being, since persons are highly future-oriented in their preferences. To kill a person is therefore, normally, to violate not just one, but a wide range of the most central and significant preferences a being can have. Very often, it will make nonsense of everything that the victim has been trying to do in the past days, months, or even years. In contrast, beings who cannot see themselves as entities with a future cannot have any preferences about their own future existence. (Singer 2011, 80)

To another extent, Singer's preference utilitarianism significantly differs from McMahan's time-relative interest account, because the latter, in recourse to Parfit's deliberations on identity and rational egoistic concern, rejects any position that ties the interest in continuing to live to questions of personal identity, as has been seen. The interest in continuing to live, McMahan argues, is *not* dependent on the ability to be conscious of oneself as a temporal continuant. Sentience is sufficient (see also chapter 3.4.1).

The question which kind of animals fulfill Locke's definition of a person, and therefore have an interest in continuing to live, cannot be answered undoubtedly, according to Singer. He is certain, however, that it *definitely* includes great apes, elephants, dolphins and some bird species and suggests to include all (or at least many) mammals and possibly also all fish and birds (Singer 2011, 119f). I will come back to Singer's problematic concept of a person below, but I would like to point out one problem that almost thrusts itself upon us here.

110 In stating that the killing of a person is *worse* than the killing of a nonperson, one might get the impression that non-persons actually do have a (weak) interest in continuing to live on Singer's account and can be harmed through their deaths as well. This is not Singer's intended interpretation, I believe, but rather owed to the circumstances that Singer acknowledges that there is no categorical distinction between nonpersons and persons. Accordingly, the wrongness of killing should not be seen "as a black and white matter, dependent on whether the being killed is or is not a person, but as a matter of degree, dependent on, among other things, whether the being killed was fully a person or was a near-person or had no self-awareness at all, the extent to which, by our best estimate, the being had future-directed desires, and how central those desires were to the being's life" (Singer 2011, 104). As Singer is also eager to include as many animals in the group of persons as possible and does not lay out the details of near-persons, it is safe to continue to assume that only persons can have an interest in their continued existence at all, according to Singer, even though two persons may considerably differ from one another regarding their abilities to have future-directed desires and their degree of self-consciousness.

Singer acknowledges that the ability to have future-directed desires is a matter of degree. He accepts that "even for those nonhuman animals who are self-aware, and hence meet our definition of 'person' it is still true that they are not likely to be nearly as much focused on the future as normal human beings are" (ibid. 103). Paradigmatic human adults typically weave their past experiences as well as their desires, hopes and dreams into a biographical narrative, for instance, and also pursue long-term goals and projects, that might stretch into the very distant future. Again, these ideas closely resemble McMahan's considerations, for instance, about psychological connectedness and narrative unity. Additionally, even though Singer does not explicitly say so, he seems to hold that certain preferences, desires etc. (i.e., direct psychological connections) may bind a person stronger to her future self than others, which would again be in line with McMahan's ideas (see chapter 3 for details).

However, Singer fails to differentiate these degrees further, but vaguely implies that "[s]ome nonhuman animals appear to conceive of themselves as distinct beings with a past and a future, and this provides a direct reason against killing them, the strength of which will vary with the degree to which the animal is capable of having desires for the future" (ibid. 119). As a result, Singer has to necessarily put different animals such as a chimpanzee and a trout under the umbrella term of a person (even though he also introduces the term "near-persons" (ibid. 104) at some point). Alas, it seems to be very implausible, to say the least, to assume that all these animals (or even only all *mammals*!) were persons in the Lockean sense. DeGrazia accordingly judges that Singer's discussion of the concept of a person got "terribly confused" (DeGrazia 1996, 242 fn 57) here.

If Singer's own claim is taken seriously, namely that only those beings, who are conscious of themselves as existing over time, have an interest in continuing to live, then all those animals without this kind of self-consciousness do not have a preference in continuing to live that must be taken into consideration. Their killing, thus, can only raise a moral problem insofar that even a preference utilitarian should be interested in the maximization of pleasure in the world and should compensate for the future pleasures that are getting lost through the killing of these animals. There can be indirect reasons against the killing of nonpersons, therefore, but no direct reasons (for details, see Singer 2011, 88ff). Thus, a preference utilitarian can still come to the conclusion that it would be wrong to kill a merely conscious animal while also denying that this would be due to the animal's interest in continuing to live. Importantly, this wrong can be made up if the killed animal is being replaced by another, similar animal, that would otherwise not have come into existence. Animals, who do not fall under the category of persons, are therefore replaceable, in Singer's account (ibid. 81, 120).[111]

Let us assume that Singer's argument is sound. Mammals, birds and fish are persons and thus possess a preference in continuing to live. The question that stills waits to be answered then is how much these animals' interests in continuing to live should be weighted. Interests can be counterbalanced by rival interests, according to the utilitarian principle of aggregation. This means that the interest in continuing to live does not imply a *right* not to be killed. Worse, from the perspective of an animal with a relatively weak interest in continuing to live speaking, it may be particularly *easy* to override her interest, if doing so resulted in a particularly useful outcome.

6.1.1 *Practical Implications for the Killing of Animals in Singer's Account*

Two areas of application, where animals are routinely being killed for human purposes, can be considered to be most pressing from an ethical perspective. The first is animal agriculture[112], which affects a huge number of animals, and the second is animal experimentation, which arguably has a very high utility. In considering the practical implications of utilitarian, rights and hybrid accounts, I will focus on these two areas hereafter.

It should be evident that a utilitarian animal ethics with its emphasis on sentience cannot justify the practice of raising, keeping and killing large numbers of pigs, cows, birds and other animals for food in factory farms. The incredible suffering and terror that industrial livestock farming inflicts on these animals has been extensively described and documented (e.g., by Singer 2009 and Foer 2009, to name just two examples), so the reader will be spared the horrifying details. Rather, it will be taken as a given that industrial livestock farming can hardly be justified from a utilitarian point of view. If it is furthermore accepted that fish are sentient animals as well and that conventional fishing causes much pain to the animals, this practice must be condemned as well (Wild 2012).

This being said, this thesis is concerned with the question how much weight an animal's interest in continuing to live should have and, thus, under which

---

111 Singer has come to defend the claim that even persons should be regarded as replaceable (Singer 2015). His replaceability argument has repeatedly been criticized (for two recent examples, see Višak 2015 and Kagan 2015.).

112 It will be presupposed that a vegetarian or vegan diet does not negatively affect the individual's health or is in conflict with any religious or cultural beliefs, and will be as easily accessible as an alternative diet containing meat. It will be furthermore assumed that the main reason for choosing an omnivore over a plant-based lifestyle is to be found in culinary preferences of the individual.

circumstances her painless killing could be morally justified. It will be assumed here and hereafter, therefore, that after having lived a happy, species-appropriate life on a farm[113], the relevant animals are quickly and absolutely painlessly killed. How much, now, should these animals' interest in continuing to live count on a preference utilitarian account?

By recourse to the principle of equal consideration of interests, Singer proposes to treat these animals just like we *de facto* treat other individuals with a comparable interest in continuing to live. As we do not raise and slaughter human infants or cognitively seriously disabled human beings, that is, individuals who plausibly do not have a (preference) interest in their continued lives at all, according to Singer, we should refrain from doing so in case of animals, many of whom do in fact have an interest in continuing to live (see, e.g., Singer 2011, 66 and 58). This kind of argument has become known as the *argument from marginal cases*, which will be critically discussed in detail in chapter 6.4.2. To anticipate one of the conclusions that will be made, Singer sometimes tends to put way too much emphasis on the *status quo* than on an honest and detailed investigation of what his theory *actually* entails and practically implies, in my view. Singer's disapproval of xenotransplantation, which will be sketched next, may illustrate this further.

The alleged utility of animal experiments has been repeatedly questioned for various reasons. It has been pointed out, for example, that the benefit and knowledge gained from these experiments may not be nearly as high as commonly thought of, or that animals are not suitable for human modeling in many cases. I believe these to be very important issues and am sympathetic to the claims that have been raised. As I am interested in whether *any* kind of animal experimentation can be justified according to the ethical theory at stake, it will be presupposed that experimentation on animals indeed strongly contributes to medical progress. Many animal experiments are accompanied by significant pain or suffering, but there are exceptions. I will focus on xenotransplantation here, as it is an extremely pressing issue in the face of scarcity

113 Even on organic farms, rearing pigs are slaughtered very young, at an average age of six to seven months, but have an average life expectancy of approximately fifteen years (Reinke 2016). Rearing pigs are genetically modified, so that their life expectancy is much lower than that of wild boars who can get as old as 21 years. In this context, the problem of specifying which of the many possible alternative longer lives should serve as the antecedent becomes evident. In chapter 3.2.1 it was argued that in the case of the rearing pig, one should compare the pig's actual short life of six or seven months with a hypothetical longer life where death would have occurred at the maximum life expectancy of rearing pigs, but not wild boars. I have also argued against the logic of the larder, claiming that as a rearing pig only exists because of the fact that it will be killed for consumption, his interest in not being killed is counterbalanced by his interest in coming into existence.

of organs – according to the DSO (*Deutsche Stiftung Organtransplantation*), approximately 10,000 patients are currently waiting for a suitable organ in Germany alone – and as it also represents a real future possibility, due to the medical progress that has been made.

Xenotransplantation describes the transfer of tissue or organs from one individual to another of a different species. Assume that a xenotransplantation from a pig would prolong the life of a 40-year-old man for another 20 years. Without the operation, the man would die within several days. In this (necessarily simplified) example a human life stands against a pig's life. According to Singer, both have an interest in continuing to live, assuming that the pig fulfills Locke's criterion of being a person. As the man's interest is much stronger than the animal's, however, Singer should come to the conclusion that the pig can be sacrificed for the sake of greater overall utility. This would even be true, if many more pigs had to be sacrificed for a single person or if the donator animal was not a pig but a chimpanzee. If, moreover, the pigs would be allowed to live a life as happy as a pig's life can be, then it is difficult to understand, why a utilitarian should not welcome the possibilities of xenotransplantation. Surprisingly, however, Singer strongly opposes xenotransplantation. In doing so, he argues by recourse to anti-speciesism:

> The idea of using animals as a source for organ donation is an example of speciesism, premised as it is on the idea that animals are things for us to use as beset suits our own interests, without much concern for the interests of the animals themselves. Perhaps the easiest way to see this is to ask yourself the following question; why should we be prepared to accept the use of organs from animals, but not be prepared to take them from human infants who are, and always will be, less intellectually developed than the nonhuman animals? (Singer 2000, 415)

6.1.2 *Some Problems with Singer's Position*

Singer continues to be one of the most popular and controversial philosophers, while at the same time attracting much criticism. A thorough discussion of these objections would go beyond the scope of this thesis, so I would like to single out one aspect that has persuaded many authors to turn towards a rights account, among them, most prominently, Tom Regan. This aspect, which may be considered to be the main challenge of utilitarian theories in general, is that Singer's preference utilitarianism cannot protect its moral objects from instrumentalization.[114] This has been exemplified above by xenotransplantation.

114 Some of the problematic implications of Cochrane's interest-based rights account, that will be sketched in chapter 6.2.5, apply to Singer's utilitarian ethics account as well.

A position that welcomes xenotransplantation on the basis of preference utilitarian considerations is coherently justifiable, but, interestingly, Singer fiercely condemns xenotransplantation. This view reveals an incoherence in Singer's ethics. By arguing that the question, whether an action is morally justifiable can be answered by asking ourselves whether we would implement the same practice on human marginal cases, he bases his prohibition of certain animal experiments on a descriptive observation, namely, that we *de facto* do not use disabled humans or infants for these experiments. It would be much more consequential to argue that, according to the principle of aggregation, there are situations where one life can (and must) be sacrificed for the life of another one. Singer actually argues against his own utilitarian principle, I believe, in implicitly arguing that there may be situations where it does not apply.

Of course, Singer must argue against the practice of raising animals with the purpose of killing them painlessly for their organs or their meat, if he does not also want to plead for the abuse of human marginal cases as donators or meat suppliers.[115] Thus, he occasionally relies on a Kantian concept of worth, as becomes evident in the below passage:

> If one is opposed to inflicting suffering on animals, but not to the painless killing of animals, one could consistently eat animals who had lived free of all suffering and been instantly, painlessly slaughtered. Yet practically and psychologically it is impossible to be consistent in one's concern for nonhuman animals while continuing to dine on them. If we are prepared to take the life of another being merely in order to satisfy our taste for a particular kind of food, then that being is no more than a *means to our end.* (Singer 2009, 159f; my emphasis)

Thus, Singer sometimes goes well beyond a simple analysis of costs and utility, if doing so will result in the kind of normative implications he wishes to defend. And indeed, considerations of the problematic implications of a utilitarian principle of aggregation regarding, among others, the practice of killing animals, caused a movement away from utilitarianism and towards deontological accounts, that grant animals a *right* not to be killed.

115 Arguments concerning the negative consequences for a society that allows abuse of disabled humans cannot be applied, because Singer does not accept slippery slope arguments, nor other arguments concerned with social stability. In contrast, the fear of becoming the victim of such an abuse oneself is relevant, in principle. As his discussion of euthanasia shows, however, Singer regards these problems to be solvable (Singer 2011, 155–59).

6.2 Animal Rights Accounts

Often it is implicitly assumed that rights must be categorical in that two individuals with the right not to be killed[116], for instance, automatically possess it equally. This is not necessarily the case, however, as will become evident. In the following paragraphs, I will first sketch Tom Regan's egalitarian abolitionist position and then turn towards Alasdair Cochrane's interest-based animal rights account. Afterwards, some objections against animal rights accounts will be addressed.

6.2.1 *Value of the Vessel: Tom Regan's Deontological Justification of Rights*

Regan published his main work *The Case for Animal Rights* in 1983, that is, eight years after Singer's *Animal Liberation* and four years after the *Practical Ethics*, and thus presented an alternative to Singer's utilitarian animal ethics account. In order to prevent the unwelcome implications of Singer's utilitarian account, which enables the sacrificing and discrimination of marginalized groups or individuals, Regan grants certain individuals an *inherent value* and argues furthermore that this inherent value justifies a right to be treated respectfully (which implies a right not to be sacrificed for the greater overall utility). This right to respectful treatment implies a right not to be killed, even though, of course, these rights are not absolute in the sense that they can *never* be disregarded in light of other moral considerations. Regan acknowledges, for instance, that if we have to choose between overriding the rights of the many or the rights of the few, then the violation of the rights of the few can be justified, of course. Similarly, situations of self-defense might make it necessary to override an individual's right not to be killed (Regan 2004, 145 and 304ff).

An individual's inherent value is not identical with and irreducible to *intrinsic values*, that is, the positive experiences a sentient being's life entails. The difference between intrinsic and inherent values can be nicely illustrated by help of Singer's metaphor of the receptacle (or vessel); while Singer attributes value only to the content (the positive experiences) of the vessel, Regan additionally ascribes an inherent value to the vessel (the individual) itself, which it possesses independently from the intrinsic value of its content. The individual thus is only valuable to Singer insofar that it contains something valuable, which means that two individuals can be more or less valuable (Singer 2011, 106). In contrast to this, two individuals with an inherent value possess it equally, according to Regan (Regan 2004, 236f).

116 The *right not to be killed* is sometimes also referred to as the *right to life*. Both terms will be used interchangeably.

Importantly, this differentiation between intrinsic and inherent values allows Regan to agree with McMahan insofar that death can do more harm to one individual than to another one, because their respective losses of future intrinsic values may greatly differ (ibid. 324). Regan could accept, thus, that one individual has a stronger interest in continuing to live than another one. This fact, however, is morally irrelevant for the ethical evaluation of their killings, at least under normal circumstances. The reason for this is that whether or not we are allowed to kill an individual depends, first and foremost, on her inherent value, justifying a right not to be killed.

In stating above that the fact that one individual will be harmed more by death than another one is *normally* irrelevant for the ethics of killing, points towards certain exceptions from the rule. These exceptional cases refer to situations, in which death will inevitably occur to one or more of the involved parties, that is, in *lifeboat cases*. In the standard scenario of the lifeboat case, there are five survivors on a boat, four paradigmatic human adults and one dog. The boat can only carry four passengers, so we must decide whom to sacrifice, if we do not want to lose all to a certain death (Regan 2004, 285). Regan now states:

> Death for the dog, [...] though a harm, is not comparable to the harm that death would be for any of the humans. To throw any one of the humans overboard, to face certain death, would be to make that individual worse-off (i.e., would cause *that* individual a greater harm) than the harm that would be done to the dog if the animal was thrown overboard. (ibid. 324)

This is an implication of Regan's *worse-off principle*[117], which authorizes the killing of the dog in this case. It is crucial to keep in mind, however, that only in extreme cases degrees of harmfulness become morally relevant at all, according to Regan. In the vast majority of cases, an animal's life is not being taken by us in situations of self-defense, but for the simple reason that doing so will have a utility for us. In ascribing a right not to be killed to animals, based on their inherent value, like Regan does, this practice becomes morally

117 "Special considerations aside, when we must decide to override the rights of the many or the rights of the few who are innocent, and when the harm faced by the few would make them worse-off than any of the many would be if any other option were chosen, then we ought to override the rights of the many" (Regan 2004, 308). For a detailed analysis of the special considerations, in which this and other principles do not (fully) apply, see Regan 2004, 315–324. Thus, even if not *one* but *one thousand* dogs had to be sacrificed for the four human persons in the boat, this action would be morally right, according to Regan, because there is no single being, in which the harm done to the thousand dogs is unified; rather, the harm that one dog suffers must be put against the harm suffered by each of the four humans (ibid. 309f; but see also 305ff).

unjustifiable. Under normal circumstances, therefore, all individuals with an inherent value have an *equal* right not to be killed.

Of particular interest now is the question which kind of animals possess such an inherent value that ascribes them with the right to respectful treatment. Regan argues that inherent value is shared by *subjects-of-a-life* who possess the "mystery of a unified psychological presence" (Regan 2004, xvi). This is reminiscent to McMahan, but in contrast to McMahan, who defines the prudential unity relations very minimally, Regan lists several conditions subjects-of-a-life must fulfill:

> [I]ndividuals are subjects-of-a-life if they have beliefs and desires; perception, memory, and a sense of the future, including their own future; an emotional life together with feelings of pleasure and pain; preference- and welfare-interests; the ability to initiate action in pursuit of their desires and goals; a psychophysical identity over time; and an individual welfare in the sense that their experiential life fares well or ill for them, logically independently of their utility for others and logically independently of their being the object of anyone else's interests. (Regan 2004, 243)

The subject-of-a-life criterion has been debated heatedly due to the fact that almost all of its capabilities are matters of degree and because it remains agnostic about animals that only fulfill some but not all conditions. Regan initially argued that at least all cognitively non-disabled mammals above one year of age were subjects-of-a-life, but later also included birds and fish (Regan 2004, xvi). This has been criticized from different angles. While, for instance, Johann Ach (2009) argues that Regan includes too many animals than his own criterion justifies, Julia Tanner (2009, 23) comes to the opposite conclusion and argues that all vertebrates and maybe some cephalopods should be involved in the group of subjects-of-a-life. Recently, Regan has dissociated himself further from the subject-of-a-life criterion in arguing that sentience should be enough to ascribe an animal with an inherent value and, thus, a right not to be killed (Regan 2007, 86–88; cited in Grimm and Wild 2016, 94f).

6.2.2 *Practical Implications for the Killing of Animals in Regan's Account*

Regan's Kantian concept of inherent value admonishes to regard those who hold it not merely as a means but as ends-in-themselves. Naturally, its practical implications are much more restrictive than Singer's preference utilitarianism, as will become evident. Regan himself calls his position a radical egalitarian animal rights account; radical, because it argues for the abolishment of the usage of animals, and egalitarian, because it ascribes equal rights to all subjects-of-a-life (Grimm and Wild 2016, 132f).

As at least all mammals, birds and fish possess an inherent value, according to Regan, their killing can only be morally justifiable under very specific and extreme conditions. The raising and killing of animals falling under the subject-of-a-life criterion, is wrong; independently of the farming type and whether or not the animals lived happy or miserable lives until their deaths. Unsurprisingly, Regan strongly argues for a moral obligation to adopt a vegan lifestyle.[118]

> *It is the impoverished view of the value of these individuals, not only the pain or suffering they are made to endure, that exposes the practice as fundamentally unjust.* Even were the individuals who were treated as renewable resources, in the sense explained, "treated well" (e.g., not caused unnecessary suffering), that would not alter the basic injustice of the practice; it would only eliminate the further wrong of harming these individuals in this way. It would make the implementation of the practice *less* wrong; but it would not remove the fundamental injustice endemic to the practice. (Regan 2004, 344)

Steven L. Davis (2003) argues that a vegan lifestyle actually results in more animal deaths than an omnivore diet containing half plants and half grass-fed cattle products. In a nutshell, the argument behind this claim is the following. A balanced plant-based diet depends on the provision of grain, legumes, soy and vegetables. Every time a field is worked, however, many small animals living or hiding in and on the ground, are being killed. Because food production for a vegan diet requires relatively intensive crop production systems, Davis estimates that it results in approximately 15 animal deaths per hectare per year. If the same land is used as pasture or for the production of animal feed, the killing of field animals could be diminished by half. He proposes, therefore, to use half the cultivated land – especially rough land unsuitable for the growing of food for human consumption – as pasture or for cattle feed production. The "Davis diet" (Lamey 2007, 332) – containing half plant-based and half grass-fed cattle products – would result in only 1.35 billion annual animal deaths, in contrast to 1.8 billion, if every single hectare in the United States was used to produce the food for a vegan diet.

118 Here and hereafter, veganism will be defined as the practice to avoid, as far as possible, all products whose manufacturing imposes any form of pain, suffering or death on sentient animals. Vegans disagree on whether a vegan lifestyle also includes the moral consideration of insects or even bacteria and whether the consumption of honey, e.g., should be condemned as exploitive. Most animal philosophers exclude insects and bacteria from the realm of moral consideration because of their alleged inability to feel pain, however, as does Regan. Thus it will be assumed that insects, bacteria and all other animals incapable of feeling pain are not proper objects of moral concern.

Davis' now famous proposal sparked a lively debate, that is still ongoing. The validity of his analysis can be disputed for several reasons, however, the most important being that a plant-based diet can feed many more consumers per hectare than a carnivore diet, and that, thus, the killing of animals per hectare must be balanced against the number of consumers one hectare feeds. Also, the high mortality rate of field killings is mainly due to predators, who take advantage of their exposed prey in a harvested field. If these deaths are left out[119], as few as 0.75 deaths per hectare remain, in contrast to Davis' 15. Under careful scrutiny, a plant-based diet actually results in much fewer deaths per consumer than the Davis diet. Regan's plea for veganism is coherent, therefore.[120]

The end never justifies the means, if it entails a violation of a right, according to Regan. The practical implications of this principle become maybe most evident in the field of animal experimentation: just as we, as a society, do not accept attempts to justify experimentation on humans by reference to

---

119 Even though it could, in principle, be argued that these deaths must be included in the mortality rate, because the cutting of the crops led to the field animals' exposure to begin with and, consequently, their deaths, this line of argumentation is extremely problematic. A careful calculation of *all* effects and consequences that the harvesting of the field has on each and every involved being, could well be too much of a challenge. The authors of the study, that Davis' analysis is based on, explicitly point out that the cutting of the crops and the corresponding exposure of a high number of prey animals is essential for the survival of numerous young owls and weasels (Tew and Macdonald 1993). What is disastrous for the mice and other field animals, can be described as a stroke of luck for other animals, thus. Lamey concludes that, "[a]ll things considered, harvesting the field might even do more good for animals than harm" (Lamey 2007, 337). Others have pointed out the positive effects in terms of sustainability that grass feeding of cows has on soil and climate (e.g., Bäuerlein 2011, Pollan 2007). One take-home message from this debate is that the question which diet actually results in less animal deaths and has more overall positive effects on soil, climate, ecosystems etc. may not be as easily answered as it seems at first glance. Consequently, it is not surprising that both vegans and proponents of sustainable meat production strongly believe to represent the globally more sustainable diet and occasionally accuse their opposite parties of naivety or a refusal to face the 'facts'. This topic is much too complex to do it justice in a footnote, however, and can be better approached by others than me.

120 For many animal ethicists it is crucial which kind of animal is killed in the particular process and at which age. Even though, e.g., Singer, as we have seen, leans towards the inclusion of all mammals and possibly birds and fish into his definition of a person, so that most killings in the field would matter morally, he still acknowledges that animal species differ considerably with regard to their cognitive architecture, which has an effect on the badness of their deaths. Other accounts also take into account the fact that meat industry kills animals at a very young age and, thus, deprives these animals of many years. Regan, in contrast, explicitly defends an *equal* inherent value to *all* subjects-of-a-life, which makes their killing equally immoral under normal circumstances.

its extraordinary overall utility for us, we should not accept it in case of animals. In spite of the fact that his rights account should permit experimentation on newly born mammals or fetuses (ibid. 398), for instance, Regan defends a total abolishment of *all* animal experimentation (for a critical discussion, see Jamieson 1990). The reason for this is that Regan aims to expand the right not to be killed to other animals as well. He justifies this expansion via a slippery-slope argumentation. Thus, Regan claims that we should rather err on the side of caution, that is, we should rather include too many animals into the realm of moral consideration than too little. Furthermore, experimentation on sentient beings, who are not subjects-of-a-life could support a "throw-away-attitude" (Regan 2004, 368) and negatively influence our relation towards those animals who certainly *are* subjects-of-a-life. (I will come back to Regan's argumentation in closer detail in chapter 6.4.)

Xenotransplantation can be regarded as a real-life scenario of the metaphor of the vessels, that has been presented above, because it actually regards animals merely as receptacles of organs for human utility. In spite of the high utility for a large number of human beings and the comfortable lives that these animals might be allowed to live until their painless deaths, Regan's rights account can never justify an action that sacrifices a subject-of-a-life by reference to the principle of aggregation. Xenotransplantation cannot be morally justified, therefore.[121]

In sum, Regan's account asks for a change towards a vegan lifestyle and a termination of animal experimentation. Before pointing towards some problems of Regan's account, Cochrane's animal rights position will be sketched. Cochrane bases rights not on the presence of an inherent value, but on interests. Therefore, his position combines elements of Singer's as well as Regan's.

6.2.3 *The Demystification of Rights: Alasdair Cochrane's Interest-based Justifications of Rights*

Cohrane's position is particularly interesting, because he explicitly refers to McMahan's time-relative interest account in determining an individual's interest

121 Xenotransplantation does not represent a lifeboat case. Because the pig was literally treated as an organ vessel, a violation of his right to respectful treatment was committed long before we came into the situation, in which one life stands against another one. As Birnbacher remarks, animal agriculture or experimentation is not offensive to Regan because of the inevitably inflictions of pain, suffering and premature deaths, that these systems entail, but, first and foremost, because beings who should be treated as ends-in-themselves, are instrumentalized (Birnbacher 2009, 56).

in continuing to live.[122] His position significantly differs from McMahan's own assessment of the ethics of killing animals, however.

Cochrane argues contra Regan that the fact that an animal *has* a right does not yet tell us anything about its strength. He argues to proceed in two steps. At the first level, *abstract* or *prima facie rights* are detected along the lines of Jospeh Raz: "'X has a right' if and only if X can have rights, and, other things being equal, an aspect of X's well-being (his interest) is a sufficient reason for holding some other person(s) to be under a duty" (Raz 1988, 166; cited in Cochrane 2012, 41f). Of course, not every interest results in a right. As Cochrane puts it, "all rights are based on interests, not [...] all interests translate into rights" (Cochrane 2012, 44; see also Simmons 2010, 378)). To result in a right, an interest must be non-trivial and possess a certain strength, that is, it must be essential to the individual's well-being. Because sentient animals have an interest in their continued lives and this interest is deemed "pressing enough" (Cochrane 2015, 210), they have a *prima facie* right not to be killed.

As these rights are only attributed *prima facie*, however, they are "defeasible and may in particular situations be outweighed by competing considerations" (ibid.). *Prima facie* or abstract rights, thus, do not yet possess what Joseph Raz calls the "peremptory force" (Raz 1988, 192; cited in Cochrane 2012, 44) of rights. They do not yet serve as a stop signal that clearly tells us how far we can go. They are abstract in the sense that they are relatively unspecific and not context-sensitive.

Moreover, even on this abstract level, the right not to be killed is not equally distributed among its possessors, due to varying degrees of the time-relative interest in continuing to live (Cochrane 2015, 206 and 210). Thus, as human persons normally have a stronger time-relative interest in their continued lives than dogs, their right not to be killed is stronger as well. In contrast to Regan, Cochrane makes the right not to be killed directly dependent on how much an individual is expected to be harmed through her death.

On the second level, these abstract rights are specified, providing concrete instructions in particular situation. Concrete rights consider competing interests of involved parties and all other relevant considerations. Among these relevant considerations are special relations persisting between the animal and the human being, for instance, a dog and her owner (see also Grimm and Wild 2016, 148–183; Garner 2013, 100–105). Cochrane acknowledges, thus, the

122 Cochrane disagrees with McMahan on how the amount of future good an individual loses should be determined. While McMahan argues from an objective perspective, Cochrane defends a subjective point of view (Cochrane 2012, 70). For details, see chapter 3.2.

moral weight of special relations. At the same time, however, he attaches only some, but not essential importance to them with regard to the attribution of rights, similar to Garner, who argues "that moral preferences based on special relations can only be seen as supplementing the moral worth of entities determined by their capacities" (Garner 2013, 101). Cochrane still argues in the tradition of moral individualism, therefore, even though he acknowledges, to a limited extent, the moral relevance of special relationships.

Cochrane's position has two central advantages over Singer's and Regan's. As a rights account, it acknowledges the value not only of the vessel's content, but the vessels themselves. This means that the violation of an individual's right not to be killed by reference to the higher net aggregate utility becomes impossible. This way, Cochrane's account can circumvent one major problematic implication of Singer's utilitarian account, thus. The second advantage stems from Cochrane's justification of rights. Rather than attempting to justify a Kantian-like inherent value to beings who do not fall under the Kantian definition of ends-in-themselves, as Regan does, he constitutes rights on the much less controversial fact that beings have certain interests that impose duties on others.

6.2.4 *Practical Implications in Cochrane's Account*

As Cochrane makes the ascription of the concrete right not to be killed dependent on the specific circumstances of a situation, he claims that *prima facie* animal rights are relatively uninformative:

> Crucially the interest-based rights approach does not provide a definitive list of animals' rights in the abstract. Instead, the concrete rights of animals are worked out in context: the interests of animals need to be identified; their strength needs to be evaluated; those interests need to be weighed against any competing interests; and they also need to be balanced against the burdens they impose on the putative duty-bearers. (Cochrane 2012, 43)

Rather than providing a theory which is then being applied to practical cases, the practical cases "are integral to the determination of the rights themselves" (ibid.). Although concrete rights must always take competing interests and considerations into account, therefore, Cochrane still makes certain general statements regarding the killing of animals for food and experimentation.

Similarly to Regan's approach, killing animals for food is morally wrong, because it violates the animals' right not to be killed. As Cochrane does not object to the use of animals *per se*, egg and dairy production may be preserved, even though both must meet very strict requirements (ibid. 86f).

Animal experimentation is morally condemned in Cochrane's account, including those experiments undertaken with the aim to cure human diseases (ibid. 205). Yet, Cochrane's account is less restrictive than Regan's in that it allows for some animal experiments, for instance, behavioral studies, because he does not defend an animal right to liberty (ibid. 52). Whereas Regan first and foremost condemns the instrumentalization of animals, Cochrane argues that there is nothing bad *per se* in using animals. Because (most) animals are not autonomous, which means in Cochrane's understanding that they do not have the "ability to frame, revise, and pursue their own conception of the good" (ibid. 11), they do not have an interest in liberty. Consequently, as the title of his main work suggests, Cochrane argues for *Animal Rights Without Liberation*. However, captivity often results in suffering, because the specific needs of the animal in question are not or cannot be met, which makes their confinement immoral after all in these cases.

Besides behavioral studies, Cochrane regards experimentation on animals permissible under the following circumstances: "If an animal born in captivity and drugged and anaesthetised so that it is caused no distress, and if that animal is not killed, then that animal's well-being is not affected. Thus, such an experiment can be deemed to be harmless" (Cochrane 2007, 314). This statement is quite surprising and is, indeed, in contrast with Cochrane's own justifications for why animals have a *prima facie* right not to be killed, as I will argue below.

In accordance with the time-relative interest account, Cochrane states that even those animals, who are only weakly psychologically continuous with themselves, "have an interest in continued life in order that they may have more pleasant experiences and greater overall well-being in their lives" (ibid. 306). Valuable experiences contribute to greater well-being and the thwarting of these does a harm. However, an anesthetized animal loses the opportunity to make pleasant experiences in the period of time, in which it is unconscious. An animal, being repeatedly drugged and experimented on in this way, should therefore be negatively affected in her well-being. If an untimely death does harm to an animal, because it forecloses many satisfactory experiences, the same should apply to cases where an animal is made unconscious over a longer period of time. If, in contrast, the animal is being drugged and anesthetized with the purpose to, say, remove a tumor, her being made unconscious is in her interest, because the operation contributes to her well-being.

To conclude, Cochrane's claim that the infliction of unconsciousness does not affect the animal's well-being and does no harm should be rejected. If he accepts that a being's sentience and her ability to make pleasant experiences is sufficient as to impose a duty on us not to kill her, then a sentient being should also have a *prima facie* right not to be made unconscious over a long period of time.

6.2.5 *Some Problems with Animal Rights Accounts*

For a long time, criticism on animal rights accounts has mainly focused on the question whether animals can be right-holders at all. As this question has been extensively discussed elsewhere (see, e.g., Cohen and Regan 2001, Grimm and Wild 2016), I will not revisit it, but presuppose that animals – just like infants – can possess rights. Another objection, which will only be touched upon, concerns an abolitionist account's occasionally extremely unintuitive consequences.

6.2.5.1 The Predator Argument

Determining a theory's concrete practical implications is, of course, a problem common to all normative accounts. The calculation of interests in a utilitarian approach can be extremely difficult, and has led some to conclude that "utilitarianism is not useful in practice, because it cannot resolve the difficult cases" (Li 2002, 600). In spite of this, the problem of determining the obligations that the ascription of rights has for us becomes especially immanent in case of rights accounts. A right is commonly assigned with a certain force that distinguishes it from an interest. Interests can be aggregated and overridden, due to their nature, but rights serve as a stop signal, forbidding any considerations about what could be gained by its violation. Despite this, however, it can be quite unclear how much an animal right really demands from us.

Animal rights theorists are quick to point towards certain exceptions for when a certain right does not apply or translate into a concrete right. For some Inuits in the arctic, for example, hunting and fishing as well as the wearing of animal fur is essential to survival, which is commonly accepted to represent an exception to the prohibition to kill animals for food or clothes, even in an radical abolitionist animal rights account (Regan 2001, 139; Cochrane 2012, 85).

There are some problems with this intuitively appealing exceptional case. Is the right not to be killed, equally possessed by all subjects-of-a-life, not even strong enough to force the Inuits to move further south (Cohen 2001, 231f)? And if not, *why*? Does the Inuit right to liberty 'trump' the animal right not to be killed here? But this would imply that animals have a weaker right not to be killed than humans (because we would, of course, never justify the same practice, if the Inuits were not hunting animals but other human beings), which is explicitly ruled out by Regan.

The problem is that the mere fact that the killing of animals is essential to some humans' survival does not make the practice itself defensible.[123,124] This is so, because the example above cannot be regarded as a lifeboat case, even though Regan seems to suggest so. A lifeboat case would be a situation, in which an Inuit got lost in the wilderness, for instance, and his only hope to survive would be to kill and eat an animal. No one, I strongly assume, would object to the violation of the animal's right in this extreme situation, where the person's survival is at stake. But why should the whole Inuit lifestyle be regarded as a lifeboat case? It is not as if the Inuits can only choose between life and death; they *can* move south, like Cohen suggested, to spare the lives of other right-holders.

Of course, these kind of considerations are rather absurd. In no way do I wish to defend the claim that Inuits should be forced to move to other regions. I hope to have shown, however, that attempts to justify the Inuits' violation of an animal's right not to be killed on reasons of self-defense, similar to a lifeboat situation, is not convincing. Cochrane might argue that the Inuits' right to liberty in fact trumps the, say, caribou's concrete right not to be killed, but he would have to further defend this claim, at the very least. Another, more promising avenue to avoid the absurd claim that the Inuits must be forced to give up their habitats would be to allude to the unforeseeable consequences that an enforced move could have regarding the stability of the group, but also the environment. This aspect has already been addressed in the David-Lamey debate in chapter 6.2.2 and I will come back to it in more detail below.

One common objection that has been raised against animal rights accounts has become known as the *predator argument*. In a nutshell, this objection states that if it is acknowledged that a certain kind of wild-living animal, a wildebeest, say, has a right not to be killed, then we as human agents should act in a way to protect the wildebeest's life against her predator, for instance, a lion. However, in case of feral animals, most of us hold a policy of non-interference. Rights theorists, thus, commonly attempt to defend this policy while also continuing to acknowledge the wildebeest's right not to be killed. Different strategies to do so have been pursued.

One attempt to do so is by recourse to moral agency, which is the strategy pursued by Regan. We as moral agents can decide what is wrong, the lion,

123 Similarly, the killing of animals may be essential to a farmer's survival *right now*, but this would be irrelevant in the face of each animal's right not to be killed. Raising these animals with the purpose to kill them later already violated their right to respectful treatment, according to Regan, and would make the practice immoral.

124 In fact, Regan himself (2004, xxxvii) makes a very similar point, albeit in another context.

sneaking up at a wildebeest, cannot. Thus, we should not correct the behavior of wild animals (Regan 2005, 357).

This, however, seems to be plainly wrong. Even if it is accepted with Regan that a lion cannot violate the wildebeest's right, this does not free us from the responsibility of interference. As Cochrane puts it: "The simple fact that children and animals are not moral agents does not mean that we should allow them to cause harm" (Cochrane 2012, 92). This becomes rather evident when imagining that the lion is sneaking up at a human person, rather than a wildebeest.[125]

Cochrane pursues another strategy to justify the claim that prey animals do not impose any obligations on us with regard to their right not to be killed. The lion kills the wildebeest with the purpose to eat her meat, which is essential for his survival. So, either we let the lion kill the wildebeest and sacrifice her, or we protect the wildebeest from being killed and, as the wildebeest is essential to the lion's survival, sacrifice the lion. Because intervention of the predator-prey relation is more burdensome than simply doing nothing, we are justified in adopting a let-them-be policy (Cochrane 2012, 94f).

This argument does not hold, I believe, because there certainly are alternatives available such as feeding the predator and protecting the prey. And are we not even *obliged*, to carry this thought to extremes, to separate wildebeests from lions and put lions on a plant-based diet? And if the costs of such an intervention are too high, then we should maybe try to minimize the number of animals living in the wild or at least not *preserve* wild animals. Furthermore, as Jamieson (1990) points out, even if the death of the wildebeest is regarded as necessary, for the sake of the argument, the pain and fear she suffers in the process of the hunt, certainly is not. It could be argued, thus, that animal right theorists should defend a duty to raise and painlessly kill these prey animals to prevent unnecessary suffering, at the very least.

Considerations of this sort are, of course, extremely unattractive. The only author to my knowledge who bites the bullet in accepting the above claims is Jeff McMahan, arguing that "[e]ven if we are not morally *required* to prevent

125 Regan himself discusses the case where the lion sneaks up at a child rather than an animal and comes to the conclusion that the reason why we are obligated to intervene in the child-case, lies in the fact that a child does not possess a "competence to get on with the business of living" (Regan 2004, xxxvii). The fact that the child cannot take care of herself is said to make all the difference here. I find this claim very unconvincing. For one thing, some prey animals, especially young, old or sick ones do not have such a competence as well. The fact that paradigmatic members of their species do, does not suffice. Moreover, I do not see why this competence should make such an important difference as to annul our duty to interfere.

suffering among animals in the wild for which we are not responsible, we do have a moral *reason* to prevent it, just as we have a general moral reason to prevent suffering among human beings that is independent both of the cause of the suffering and of our relation to the victims" (McMahan 2010). The prospect of giving up a policy of non-interference with regard to feral animals provokes a bad feeling in many, however. The reasons for this unease are commonly to be found – besides potential religious beliefs of a prohibition to tamper with nature or to play god – in consequentialist considerations of the following kind, that have repeatedly been brought up above.

Interferences with ecosystems regularly have unforeseen, negative outcomes. We cannot be sure what kind of damage our tampering with predator-prey relations will cause in the long run. Thus, it should be best to not intervene at all. Cochrane agrees that this sort of consideration is valid from a total perspective of utilitarianism – that is, the perspective of the universe. Rights accounts, however, should be concerned with individual well-being and reject the total version of utilitarianism. This is the reason, after all, why rights theorists typically object to the utilitarian principle of offsetting rivaling interests against one another, so that the interests of few can be sacrificed for the overall utility. Thus, Cochrane states, a prey animal with a right not to be killed must be protected, *even if* this would negatively affect the overall well-being, especially in cases like this one, where the outcome is extremely difficult to calculate so that it only remains a vague possibility that our interference would actually turn out to be a mistake (Cochrane 2012, 93).

In contrast to Cochrane, I believe that references to unforeseen negative consequences indeed are the best strategy to justify a policy of non-interference in a rights account. In cases where the consequences of our actions are so unpredictable yet potentially severe, it can be best to rather do nothing. We are just beginning to learn the complexity of ecosystems. There are numerous examples from the past, where the best intentions, for instance, the deliberate introduction of certain plants or animals, had unforeseen, sometimes devastating, consequences. For instance, the reintroduction of wolves in the American Yellowstone National Park in the mid-1990s led not only to a shift in elk and other animal populations, but in fact re-shaped the park's forests, rivers and landscape. Thus, the reintroduction of an animal species had a strong and lasting effect not only on the its prey or rivals, but on the park's whole ecosystem. The effects were mainly positive in this example, but it illustrates how unpredictable the results of any kind of interference with an ecosystem can be.

Rights accounts can accommodate this consequentialist fact, because they are not ignorant to the effects of their actions, of course. On the downside, this might involve the need to accept that "it would be instrumentally good if predatory animal species were to become extinct and be replaced by new herbivorous species, provided that this could occur without ecological upheaval involving more harm than would be prevented by the end of predation" (McMahan 2010). This means that the main reason for not bringing predatory species to extinction is our present ignorance of the potentially negative outcomes of this intervention. From this it seems to follow also, however, that we should strive to overcome this ignorance with the purpose to finally end an unjust situation in the wilderness. For most of us, this will be an absolutely unpalatable solution, of course, and indeed restore the unease that the prospect of systematic interferences into stable ecosystems provoked.

To conclude, for those animal rights accounts like Cochrane's and Regan's arguing in the tradition of moral individualism and rejecting a relational position – which would allow them to acknowledge that we are morally obliged to protect the lives of our pets but not of feral animals, for instance – the predator argument indeed poses a challenge.

6.2.5.2 The Moral Significance of Intrinsic Values

It has been criticized (e.g., by Pluhar 1988, Jamieson 1990, Edwards 1993, Garner 2013, 175, fn 8) that Regan's reliance on intrinsic values is inconsistent. Even though all passengers in the lifeboat have equal inherent value and, therefore, the equal right not to be killed, death does more harm to each one of the person than it does to the dog, according to Regan. This reference to the amount of future loss is not justified, however, in a theory that assigns rights independently from the interest in continued life, as, for instance, Rem Edwards argues: "Equal rights supposedly depend on equality of inherent worth, but in the lifeboat situation inequalities of intrinsic worth give the person stronger rights to life and freedom from harm" (Edwards 1993, 234). Edwards (as well as Pluhar 1988) argues, that for reasons of consistency, Regan should rather approve of a scenario where the decision whom to sacrifice is based on mere luck:

> [A] lottery like drawing straws is precisely the correct moral strategy for choosing among equals when lifesaving resources are too scarce to save all. Regan rejects a lottery only at the price of ignoring his own cherished doctrine of equality of inherent worth among subjects-of-a-life. In the final analysis, on Regan's view, all that we need to consider is unequal intrinsic worth. (Edwards 1993, 234)

Under careful scrutiny, I believe that Edward's criticism does not hold. An equal right of two individuals not to be killed can be defended on the grounds that they both have a strong interest in continuing to live, while also defending the claim that in lifeboat cases, but *only* in lifeboat cases, where at least one individual will necessarily lose her life, the particular strength of each individual's interest matters. Basing one's decision on how much each one of the passengers would *de facto* lose from death, would be rational in these cases. This is similar to the difference between death occurring earlier or later in the life of one and the same individual. One could argue that at both points in time, the individual has an equal right not to be killed, while also accepting that her interest in continuing to live may become much weaker.

It is, of course, another question whether a person and a dog, whose interests in continuing to live vary so extremely, *should* be assigned an equal right not to be killed to begin with. Regan's mysterious attempts to justify the egalitarian right not to be killed on a criterion, whose elements all are matters of degree, indeed initiated Cochrane's turn towards an interest-based rights account.

6.2.5.3 Justified Violations against an Animal's Right Not to be Killed

Cochrane argues that whether or not a being's *prima facie* right not to be killed translates into a concrete right in a specific situation, also depends on contrasting interests and other considerations that might stand against the individual's interest in continuing to live. The interest in continuing to live, thus, can be overridden in concrete cases. His account, thus, might be argued to be in danger of losing the peremptory force that rights accounts should have.

This becomes particularly evident in one of Cochrane's own examples, in which a large number of rats takes up residence in one's home. Imagine that life trapping or any other way to drive the rodents off is unsuccessful. Each rodent has a right not to be killed, according to Cochrane, and this right can only be "overridden for compelling reasons" (Cochrane 2015, 213). But what exactly *is* a compelling reason? Cochrane, in fact, states that the relatively weak interest of the rats in continuing to live can be overridden by our interest in avoiding diseases that may potentially be transferred from rats to humans. Even though the probability of dying of such a disease is extremely low, the rats' interest in continuing to live is so weak as to justify their killing, if no other nonlethal method is available (ibid.). This might evoke the impression that Cochrane's account is all about utilitarian consideration of interests, in the end. Why talk about rights at all?

Cochrane would probably respond by pointing out that he still argues in opposition to the utilitarian principle of aggregation. An animal thus cannot be sacrificed by reference to the higher overall utility. Importantly, this means that if there are alternatives that allowed us to save the rats' lives, then these alternatives are morally binding. It is an implication of the concept of rights that it offers protection against systematic discrimination. Consequently, Cochrane (2015) compares the situations where the right not to be killed can be offset with cases of self-defense. Therefore, it should be clear that even to *consider* violating an animal's right not to be killed can only be permissible in extreme scenarios. A different way to put this is to say that Cochrane reverses the onus of proof: someone wanting to act against another one's interest in continuing to live must provide coherent evidence for why this option represents his only just and reasonable course of action.

Assume, in a modified version of the rodent-example above, that feeding the rats a contraception represented a possible – but very expensive and extremely tedious – solution to the problem. One after another, the members of the rat family would die until finally no animal would be left. In contrast, the painless killing of the rodents would be quick, effortless and inexpensive. Would it still be justified to choose to violate the rodent's right not to be killed, according to Cochrane?

I stipulate that the higher costs of the contraceptives alone would not suffice to justify the killing of the rats. If, however, the costs for the contraceptives would put me in severe financial distress, for example, the killing of the rats may be justified. It should be evident by now, however, that Cochrane himself or other people arguing along an interest-based rights account might come to completely different results. Similarly to utilitarian calculations, therefore, there will be disagreement concerning the question how much exactly we are obliged to do in respecting an animal's right not to be killed in concrete situations. That the interest in continuing to live of right-holders can be overridden at all, makes Cochrane's account vulnerable to losing the special force immanent to rights accounts; especially in situations where we as human agents – forced into the role of the neutral judge and the partial lobbyist at the same time – have a particularly strong interest in the death of the involved beings.

6.3 Hybrid Accounts

The term *hybrid accounts* will refer to theories including consequentialist as well as deontological aspects. Cochrane's position may be argued to be hybrid insofar that it bases rights on an individual's corresponding interest and not, like Regan, on an deontological inherent value. Still, his account will be classified as a rights account here, because sufficiently strong interests are translated into rights. Hybrid accounts argue, in contrast, that interests do not necessarily transfer into rights, not even those, that impose moral duties on moral agents. Consequently, an interest in continuing to live will not necessarily result in a right not to be killed on these accounts, because the ascription of a right is argued to depend on certain other conditions going beyond the being's mere interest in continuing to live. As will become evident below, both Robert Garner as well as Jeff McMahan emphasis the moral relevance of autonomy in this context.

I will first outline Robert Garner's hybrid account and afterwards represent McMahan's own suggestion for how the time-relative interest account should be translated into a normative theory. Analogously to the discussion of Singer's preference utilitarianism as well as Regan's and Cochrane's rights accounts, some (allegedly) problematic implications will then be alluded to.

6.3.1 *Robert Garner's Qualified Rights Account*

Like Cochrane, Robert Garner argues that Regan's radical egalitarian position does not do justice to the fact that animals and persons differ very much with regard to the strengths of their interests in continuing to live and in liberty. Both agree that if the interest in continuing to live can be held to so various degrees, it makes no sense to ascribe an equal right not to be killed to all beings with this interest. Following Singer's principle of the equal consideration of (equal) interests, Garner judges that "adopting an animal rights position does not require us to justify moral equality between humans and animals. Instead, the like interests of humans and animals are to be treated equally, and their unlike interests are to be treated unequally" (Garner 2015, 224).

Because Garner implicitly defends an egalitarian view of rights in that he assumes that if two subjects have a right, they necessarily possess it equally, and because persons and nonpersons differ so immensely from one another, Garner believes that only humans have a right not to be killed, even though animals have an interest in their continued existence.[126] As we have already seen with Cochrane, this non-gradual concept of rights is not universally shared by rights theorists, however.

Personhood, probably best understood here as a combination of various cognitive abilities, plays a central role in Garner's theory, thus, because it separates those subjects with a right not to be killed and those without. In Garner's

126 Garner wants to preserve the expansion of the right not to be killed to *all* human beings, i.e., infants, cognitively severely disabled human beings, and other humans that are also referred to as 'marginal cases'. Even though Garner judges the argument of marginal cases to be "correct in theory" (Garner 2015, 220), he believes that it has and will never have any political force in that people will never be persuaded by the argument and stick to their speciesistic intuitions. As Garner defends a move away from "the rationalism typical of contemporary Western philosophy" with its emphasis on logical consistency and precise argumentation and towards "social and political realism" (ibid. 219), accepting that some things like speciesism will remain unalterable in practice, he believes that the argument of marginal cases, despite its logical correctness, may have to be abandoned. I will come back to the argument of marginal cases in more detail in chapter 6.4.

understanding, personhood entails rationality, self-consciousness as well as "a sophisticated communication system" and "moral agency" or autonomy (Garner 2015, 218; Garner 2013, 129). Especially autonomy, understood as the ability to deliberately choose one's goals and structuring one's life in a way as to pursue them, plays a major role with regard to the right not to be killed in Garner's account.

Because animals are living entirely in the present, according to Garner, and do not possess autonomy, they have an interest in their continued lives in the sense that they are deprived of the opportunities to experience pleasure, but not in the sense that their efforts or desires are being thwarted. Hence, their lack of autonomy and other capabilities associated with personhood justifies the threshold between those beings with a right not to be killed and those with only an interest in their continued lives. Because the animal and human interests in the avoidance of pain are so comparable as to demand equal consideration, Garner acknowledges that animals have "a right not to suffer from the hands of humans" (ibid. 225), in contrast.

Garner's account is a hybrid account, as defined here, because even a sufficient interest does not always result in a right for him, in contrast to the view that Cochrane defends. Both Garner and Cochrane agree that if an interest manifests itself so differently in two individuals like the interest in continuing to live, then it does not make sense to say that both have the equal right not to be killed. Whereas Cochrane concludes from this that the right not to be killed must be distributed unequally, Garner, arguing for a non-gradual conception of rights, introduces a threshold that separates those individuals with a right not to be killed, that is, persons, from those with merely an interest in continuing to live, that is, nonpersons.

6.3.2 *Practical Implications for the Killing of Animals in Garner's Account*

Individuals with a right not to be killed, that is, persons, must not be sacrificed by reference to a greater good. Individuals with an interest in their continued lives, but no equivalent right not to be killed, in contrast, can be sacrificed. Thus, Garner's practical implications regarding interest-bearers are very similar to Singer's and regarding right-holders very similar to Regan's. Animals have a right not to be made to suffer, so industrial livestock farming cannot be morally justified on Garner's account, as it inflicts tremendous pain and suffering on the animals. Because animals, however, do not have a right not to be killed, nor a right to liberty, they are no means-in-themselves and can be used, in principle. This means here that if the utility that could be gained from an animal's death is so great as to override the animal's interest in continuing to live, her killing is morally permissible.

Yet, Garner also states that "only very significant human benefits (such as the protection of human lives) would have to accrue from the loss of animal lives at human hands" (Garner 2015, 225). Garner indeed repeatedly emphasizes that an animal's interest in continuing to live can only be overridden in situations, when human lives are at stake (ibid.). Clearly, the human gustatory interest in meat is no such case. For this reason, Garner believes that the killing of animals for food cannot be morally justified, because the human interest in eating meat does not suffice to override the animal's interest in continuing to live.

Garner's conclusions can be questioned, however. In fact, it is not immanently evident why relatively *moderate* benefits should not suffice to override an individual's interest in continuing to live in at least some cases. As Garner acknowledges that the interest in continuing to live can be very weak in some subjects, he should also accept that comparably *weak* contrasting interests may suffice to offset this interest. Garner, just like McMahan, also assumes that the strength of the interest in continuing to live greatly depends, among other things, on the said animal's cognitive capacities. To be consistent, Garner should hence accept that a frog's interest in continuing to live, for instance, can be much more easily overridden than the corresponding interest of a chimpanzee. It could be questioned, thus, whether Garner's claim that *any* animal's interest in continuing to live can only be overridden in extreme cases can be upheld in his own theory. (I will get into greater detail of this aspect below, when discussing the practical implications McMahan argues his theory to have.)

As Garner differentiates between an ideal and a non-ideal moral theory, where the former describes what animals are owed as a matter of justice, while the latter represents the more practical route how this ideal can be achieved on the long run, he does not call for a radical and immediate abolition of meat industries. Instead, he argues that free-range farming and in-vitro meat production could provide morally permissible ways to reach these goals. It is important to note, however, that these options are only justified, if they do not violate the animals' right not to be made to suffer. Under the assumption that genetically engineered animals, whose capability to feel pain had been blocked, continued to suffer from distress, anxiety and other negative experiences in factory farms, this research avenue is argued to be illegitimate, therefore (Garner 2013, 136).

The non-consideration of an animal's interest in continuing to live is justified, Garner argues, when human lives must be protected. Even though this sounds similar to Regan's discussion of killing in self-defense, Garner interprets this much broader. He allows for no kind of animal experimentation involving the infliction of pain or distress (Garner 2013, 135f). He does not object to the use of animals *per se*, however, nor does he defend an animal right not to be killed. Thus, killing animals "for the development of life-saving medical treatments" (Garner 2015, 225) is permissible. (Note that in emphasizing the *development* of life-saving medication, Garner, in fact, leaves it open whether basic research represents a sufficient reason for the conduction of animal experimentation as well.) Tissue experiments, for which animals are raised and killed to generate tissue that is then experimented on, are ethically unproblematic on this account (Garner 2013, 137). For similar reasons, Garner should not be opposed to xenotransplantation as well.

6.3.3 *Jeff McMahan's Two-tiered Account*

It might come as a surprise to learn that Jeff McMahan argues for a hybrid account, because one might expect him to defend a utilitarian *time-relative interest account of killing* evaluating the wrongness of a particular act of killing according to the degree of harm the victim suffers through the violation of her time-relative interest in continuing to live. As will become evident in the following paragraphs, McMahan's considerations for why such an account alone is insufficient for a universal ethics of killing are very similar to Garner's. Both agree that animals do have an interest in their continued lives and that this interest imposes moral obligations on us. Because persons, however, are defined through certain cognitive abilities that make them so fundamentally different from nonpersons, their killing raises moral problems that go beyond the mere consideration of their time-relative interests, both argue. Below, McMahan's reasons for including rights into his account will be outlined.

As a time-relative interest in continuing to live varies not only with the degree of psychological relatedness, but also with the amount of future good, death does not only generally do more harm to a person than it does to a dog, but also to one person in contrast to another one. According to the time-relative interest account of killing, this would imply that the killing of one person is morally more repulsive than the killing of another. This, however, seems a gross violation of our egalitarian sense of the moral worth of all persons, which McMahan calls the *equal wrongness thesis*. This intuition is too strong, McMahan states, as to give it up lightly.

Of course, there certainly are situations where the killing of one person is considered to be worse than the killing of another one. Examples may be cases in which one person wishes to die, but the other person expresses her strong desire to continue living, or in lifeboat scenarios, where we must choose to either sacrifice an old or a young person. Thus, even Regan could accept that when we are forced to decide between two lives, it is fully acceptable, even from the perspective of a rights account, to make this decision directly dependent on how strongly they will be harmed through their deaths. To determine in a time-relative interest account how strong two persons will be harmed by their deaths also requires considering their respective *ways of living*, because how a person will lead her future life will affect the amount of future good she will be expected to lose from dying.

Importantly, however, we do not tend to think that how two persons live their lives, that is, their respective *ways of living*, should be morally relevant under normal circumstances, that is, no lifeboat or other extreme situations.[127] To illustrate this, imagine the following two individuals, that have been introduced by Evelyn Pluhar:

> Individual A will never again engage in his favorite activities, such as playing the piano, reading Dostoevsky, programming his computer, listening to Beethoven's Ninth Symphony, eating delicately seasoned Indian cuisine, or meditating in a Japanese tea garden. B, by contrast, loses his regularly scheduled pleasures of drinking Iron City beer, watching "Celebrity Wrestling" on TV, and going bowling. (Pluhar 1988, 215)

Even though McMahan, in contrast to Pluhar, should accept that A will lose more from death than B, he rejects the idea of evaluating A's *killing* worse. The reason for this lies in the fact, he stipulates, that all persons possess an equally strong worth that naturally comes with them being persons and that lies at the basis of why we think that it is generally equally bad to kill two persons. This intuition is sufficient to justify the move from interests to rights, according to McMahan; in fact, rights for him come into play *only* in situations, where the interest account would have counterintuitive conclusions:

> An appeal to rights is necessary only when a principle requiring respect for interests cannot account for the moral reasons we seem intuitively to have. And these reasons seem to arise only in our dealings with individuals that have not

127 Pluhar, in fact, argues that two individuals' ways of living *never* should be considered as morally relevant, i.e., not even in lifeboat cases: "While an individual who is merely shot [in the leg] is better off than one who is killed, just as the same individual is better off being merely shot than killed, two individuals who are killed are equally badly off in that each *loses everything*" (Pluhar 1988, 215). On the one hand, she thus disagrees with Regan, who argues that two individuals with equal inherent worth can be harmed to different extents by their deaths. On the other hand, Pluhar agrees with Regan insofar that age and well-being can be morally relevant regarding the harm of death in that death of a young or a healthy individual is generally worse than death of an old or terminally ill (ibid. 216). She considers these to be only exceptional cases, however.

> only interests but also certain higher cognitive and emotional capacities, such as self-consciousness, autonomy, and rationality. (McMahan 2008, 70)[128]

If Cochrane's interest-based rights account demystifies rights, McMahan, in a sense, re-mystifies them, because he establishes that some, but not all, individuals have an inherent worth or value, just like Regan did. While Regan argues, however, that this inherent value is equal to all subjects-of-a-life (or sentient beings, in more recent writings), McMahan restricts it to persons. Importantly, McMahan's concept of persons is much more demanding than Singer's, however, because he closely ties personhood to the capability of autonomy, which he defines as "the capacity to direct one's life in accordance with values that one reflectively endorses" (McMahan 2002, 256).

McMahan now proposes a threshold separating autonomous beings, that is, persons, from non-autonomous beings, that is, nonpersons. The morality of killing above the threshold considers all killings to be equally bad under normal circumstances, regardless of the respective harms done through death, while the killing of two individuals under the threshold is to be evaluated according to the harm suffered through death. The resulting model is a two-tiered account, consisting of a moral theory of respect or rights above the threshold, and a time-relative interest account of killing for all sentient beings underneath the threshold. This resembles Robert Nozick's (1974/2008) famous formula 'utilitarianism for animals, Kantianism for people'.

As McMahan himself mentions, the postulation of a such a threshold however, suspiciously looks like "an arbitrary, ad hoc stipulation motivated entirely by a desire to salvage our egalitarian intuitions" (McMahan 2002, 249). If he accepts that all sentient beings have an interest in their continued existence, then why do some but not all have an equal right not to be killed? How can it be justified to separate those beings with a certain cognitive capability from those without? Unfortunately, McMahan does not go into further details of this topic, but I assume that he bases his claims on a consideration of respect for a person's autonomous decisions.

Pluhar believes that if we want to preserve the equal wrongness thesis and thus our intuition that to either kill A or B is equally bad, regardless of their respective ways of living, then the same should apply if B is a dog. But is it really true that the two cases are analogous? B *could* go to the theater, in principle, make a computer course, and learn to play the piano. He just does not want to.

128 A justification of rights entirely on the basis of certain intuitions is highly questionable, I believe, but for lack of space the details of this problematic aspect cannot be laid out here. Instead, I will focus on McMahan's move from interests to rights.

He autonomously chose a different life and also different values than A. As a person, B is able to evaluate his possibilities and is free to do whatever he wants to.[129] So, if we came to the decision that killing B would not be as wrong as killing A, we fail to respect B's autonomous decision, it could be argued. In case of the dog, this argument may not be transferable, however, because the dog did not autonomously choose his way of living, that is, he did not deliberately decide to pursue certain values or long-term projects.

I stipulate that considerations of this kind lie at the basis of why McMahan judges autonomy to be of vital significance for the moral evaluation of a killing. "[T]he autonomous determinations of [a] person's will" (McMahan 2002, 260), demands respect, he argues, and seems to be a significant element of his intuition that some beings, that is, persons, possess an equal inherent value. This value, in turn, justifies them having a higher moral status than other beings and, importantly, guarantees them equal moral status above the threshold.

The capacity of autonomy as defined by McMahan involves a set of psychological capacities coming in degree, such as self-consciousness, rationality and agency. Again, this reminds one of the objections that have been raised against Regan's subject-of-a-life criterion; it seems bizarre, to say the least, to represent the obtaining of autonomy as an abrupt transition over the threshold, even more so, as the ethical consequences are so serious. How is the killing of a 4-year-old child possessing some but not all of the capacities that are involved in the evolvement of autonomy to be evaluated on this account? If the child is to be considered as a member of the class of beings above the threshold, his killing is as equally bad as the killing of an adult. If he is to be classified as a nonperson belonging beyond the threshold, this will necessarily lead to a very different result.

These considerations, together with the equal wrongness thesis, suggest a modification of the two-tiered account. One could either raise the moral status of all those beings moving in the gray area of indeterminacy on the basis of a slippery slope argument, or one could broaden the two-tiered account by introducing a third, intermediate stage. The latter position, which McMahan favors, regards the threshold not as a sharp line but as "a broad band of indeterminacy" (McMahan 2002, 265) that includes all beings possessing what could

129 I am fully aware that this represents a very idealistic, and potentially unrealistic depiction of autonomous decision-making. The important aspect here is that persons generally feel responsible for the outcome of their lives and in control of what they pursue and desire. This is also related to the capability of event-independent thinking, because it enables subjects to regret decisions they know they had been or will be able to make differently (see chapter 4.2.4).

be described as the 'rudiments' of autonomy. He acknowledges that this will plausibly not only encompass infants but also some animals:

> Certain animals – the higher primates, for example – may also fall within the area of indeterminacy and thus share the intermediate status of children in the process of becoming autonomous. This, I believe, is a welcome result. It seems plausible to suppose that the status of the higher primates is intermediate between that of lower animals that altogether lack self-consciousness and autonomy and that of persons. (McMahan 2002, 265)

McMahan believes that the killing of those animals on the intermediate stage should be determined neither through a rights account, nor through a time-relative interest account of killing. Rather, he suggests that this intermediate stage probably makes a third normative theory necessary, which he does not further specify.

6.3.4 *Practical Implications in McMahan's Account*

In contrast to Garner, who acknowledges an animal right not to be made to suffer, McMahan completely excludes animals from the realm of rights. If competing interests are so strong as to outweigh an animal's interest in continuing to live, an animal may be killed. As, according to McMahan, the net amount of good an animal loses is comparatively small and the psychological relations that persist between the animal now and the animal in the future are very weak, animals only have a weak interest in continuing to live. Consequently, it should be easily outweighed. Somehow surprisingly, however, McMahan insists on relatively strict moral obligations. Because McMahan has held a central position throughout this thesis, the following remarks will be more detailed than the remarks on previous authors' practical implications. I will in fact argue that McMahan does not coherently apply his own theory to the practical cases he discusses. Some objections against McMahan's two-tiered theory itself will be discussed below.

McMahan confidently claims that even "benign carnivorism" (McMahan 2008, 66), that is, the consumption of humanely raised and painlessly killed animals, cannot be justified under the prevailing circumstances with regard to sentient animals, such as mammals, birds and fish, but not shrimps or insects.

I will argue that the TRIA may not be able to defend such a strict obligation to vegetarianism, however. McMahan's determination of the animal's time-relative interest in continuing to live is much too strong and seems indeed to be at odds with his own theory. McMahan explicitly states that in order to find out whether a certain action involving the use of animals is permissible, we must "weigh[...] their interest against ours, or, more generally, against *whatever interests conflict with theirs*" (McMahan 2005, 531; my emphasis). It is not comprehensible why an animal's interest in not being killed for food, which can *at best* be only weak compared to persons, cannot be offset by competing, equally weak human interests. (I made a similar point with regard to Garner's position, so a great part of the following argument will apply to his theory as well.)

The strength of a time-relative interest in continuing to live varies with the net amount of future good. The older an animal is, therefore, the less objectionable it should normally be to kill her, even more so, if the animal in question is only very weakly bound to her future self. So, if it is assumed that one large, only very weakly psychologically unified animal, having happily lived several years and being painlessly killed, which could provide meals for twenty human 'gourmets' who would greatly enjoy these meals – why should that not be enough to make this act permissible? Or, to put it more generally: how much is enough to override an animal's interest in continuing to live? Just like other utilitarian theories, the TRIA is at a constant risk of manipulating the calculation in such a way as to rationalize certain already held convictions (see also Singer's dubious rejection of xenotransplantation).

This problem becomes particularly evident in McMahan's disapproval of egg consumption. Because the production of eggs normally comes along with the killing of male newborn chicks, its production, McMahan argues in an interview with Joshi Gottlieb (2014), is immoral. This conclusion is surprising, however, because birds are only very weakly psychologically connected to their future selves, according to McMahan's own assessment, and, importantly, newborn chicks still much less. McMahan's statement becomes even more surprising if put in direct comparison to his claims regarding developed human fetuses (and newborns):

> The developed fetus cannot envisage or contemplate its future and hence cannot have future-directed psychological states, such as intentions; [...] it now has no psychological architecture – no beliefs, desires, or dispositions of character – to carry forward into the future. It is [...] isolated from itself in the future. Its future is, figuratively speaking, relevantly like someone else's future. (McMahan 2002, 275f)

Such a being is only very weakly psychologically connected to and continuous with his future self. Thus, McMahan goes on:

> [I]mmediately after the developed fetus begins to exist[130] and for some time thereafter, its mental life, such as it is, is confined entirely to the present; its

130 that is, between 20 and 28 weeks after conception, according to McMahan

> mental states do not refer forward or backward in time and there are no elements of its mental life, such as beliefs or desires, that persist over time. It is psychologically locked in the present. It has no psychological connections whatever to itself in the future. […] it seems reasonable to conclude that the developed fetus's time-relative interest in continuing to live is *extremely weak*, despite the amount of good its future can be expected to contain. According to the Time-Relative Interest Account, therefore, the moral objection to killing the developed fetus immediately after it begins to exist should be *very weak*. (ibid. 277; my emphasis)

Because a chick only loses a *fraction* of the amount of good that a normal human fetus loses through her death, his time-relative interest in continuing to live is thus even significantly weaker than "extremely weak", according to the TRIA. Crucially, McMahan assumes indeed that their psychologically diminished existence state does not even change much through adulthood:

> *Like ourselves in the earliest moments of our lives*, most animals are, throughout their lives, largely psychologically *unconnected* to themselves in the future. They live mostly in the present. So not only is the life they lose through death inferior in quality and quantity, but they are also only weakly related to their possible future life in the ways that matter. (McMahan 2015, 71; my emphasis)

Recall from above that in order to determine whether an action is morally permissible, McMahan asks us to "weigh their [i.e., animals'] interest against ours, or, more generally, against whatever interests conflict with theirs" (McMahan 2005, 531). In contrast to this, McMahan often tends to give the impression as if only very *significant* human interests are to be considered in this context, for instance in life-threatening situations. Compare this to the following quotation, in which McMahan seeks to argue against the argument that an abolishment of the meat-industry would threaten the lives of many people:

> The problem with this argument is that it can be advanced on behalf of any large-scale social practice, however iniquitous. The same claim has been made in support of slavery in the South, the sale of tobacco products at home and abroad, and the sale of advanced weapons to tyrannous and illegitimate governments. It is inevitable that there will be resistance to phasing out an *immoral* practice from those whose livelihoods depend on its continuation. (McMahan 2005, 533; my emphasis)

A utilitarian account like McMahan's rivaling interests against one another in order to *figure out* which kind of action is the morally right one to begin with, must not arbitrarily take certain interests out of the calculation in regarding them as mere obstacles that must be overcome on the way to implementing what is the morally right thing to do. In fact, the examples he gives in the quotation above are rather inadequate, because they concern the *rights* of human persons, and are not applicable to animals, who are, according to McMahan's approach, excluded from the realm of rights. Chicks do not have a right not to be killed, nor do they have an only moderately strong interest in continuing to live. If, therefore, even the "moral objection to killing the developed fetus immediately after it begins to exist" (McMahan 2002, 277) is only very weak, then the moral objection to killing the hatched chick should be almost negligible. It is not comprehensible, therefore, why McMahan strongly opposes to the systematic killing of male chicks that comes along with today's mass production of eggs.

To sum up, the more than weak interest in continuing to live of a male chick, that will have to be killed for every other egg we eat, should be able to be overridden by a nearly similarly extremely weak human interest. If it is accepted that, first, some people might have a stronger interest in eating a particular type of food than others, and, second, that other (economic) considerations in favor of egg production need to be taken into account as well, then the human interests in egg production and consumption could in fact outcompete the animal's interest in continuing to live. In conclusion, McMahan's TRIA may not be able to rationally oppose to benign carnivorism, at least in case of very large, psychologically nearly unconnected animals. Furthermore, it is not easily comprehensible why McMahan should reject the killing of newly hatched chicks.[131] (This is not to say, of course, that the routinely and systematic killing of male chicks, only minutes after they have hatched, can never be morally condemned, of course.)

With regard to animal experimentation, McMahan argues that if the benefits are very great, the killing of animals is morally permissible. Xenotransplantation may provide one such example. Especially in cases where the benefit for humans will only be *potentially* available at some point in the future, for instance, in basic research, an experiment's utility may be particularly hard to calculate, however. Here, McMahan suggests a sort of test, which is, once again, based on the argument of marginal cases and on intuition.

---

131 A possible response could be that the time-relative interest in continuing to live *per se* is stronger than any other interests or considerations, so that it has more moral relevance in a utilitarian calculation. This, however, would either demand a modification of McMahan's concept of time-relative interests or a specification of the exact circumstances, under which a time-relative interest in continuing to live can be rightfully overridden at all. As has become evident in Garner's account, however, it is not sufficient, I believe, to simply identify these circumstances without providing an explanation for why even extremely weak interests in continuing to live should suffice to demand such high moral obligations on behalf of human agents.

To find out whether the potential utility of an experiment is sufficiently important to justify the killing of an animal, the experimenter should ask himself whether he would be willing to conduct the same experiment on a severely cognitively disabled infant, orphaned at birth (McMahan 2005, 535). If the experimenter would be unwilling to use an infant, his willingness to use "an animal with equivalent or even higher psychological capacities" would be at least "entitled to demand a compelling explanation of the difference" (ibid. 536). As, however, McMahan thinks that there is no such difference, and the experimenter, thus, would be unable to give a sufficient explanation, the use of animals but not infants for experiments is unjustifiable.

I believe that this line of argumentation is dangerously flawed. The major problem lies, once again, in the decisive role intuitions are allowed to play here. First and foremost, it is very plausible to assume that intuitions vary. A scientist with a mentally disabled nephew might have very different intuitions concerning the orphaned, mentally disabled child than a scientist who has never formed a strong relation to a cognitively disabled human being before. (It is even plausible to assume that the scientist's cultural background is relevant; a German scientist probably has much greater reluctance against using mentally disabled humans for experiments than someone not familiar with the *Aktion T4*, in which over 70,000 physically and cognitively disabled humans were killed between 1940 und 1945.) To make the question what to do in these cases solely dependent on one's intuition is unacceptable, in my opinion. Also, the argument of marginal cases is rather problematic in itself, as will be argued in more detail below. I will thus come back to McMahan's argument regarding the justifiability of animal experimentation in chapter 6.4.2.

6.3.5 *Some Problems with Hybrid Accounts*

6.3.5.1 Do Rights Always Trump Interests?

One central challenge of hybrid accounts is to provide an answer to the question how interests and rights are to be aggregated. How much weight has the relatively high interest in continuing to live of individual A in contrast to a rather moderate right to liberty of individual B, for example? Can interests ever trump rights?

A person's right to liberty is severely affected by the rats that have invaded her house, to draw on Cochrane's example once more. On the one hand, Garner as well as McMahan would want to emphasize that the rats' interest in continuing to live is extremely weak, which might give the impression that the person's right to liberty easily offsets the rats' weak interest. On the other hand, Garner and McMahan seem to hold that even an extremely weak interest in continuing to live can only be overridden in very exceptional cases, where human lives are at stake. Consequently, they sometimes argue that an animal's weak interest in continuing to live can still inflict serious moral obligations on us. In a nutshell, Garner as well as McMahan argue that animals have an interest in continuing to live, but it often remains unclear what exactly this entails and how this interest should be weighed against other interests and considerations and, first and foremost, other rights.

6.3.5.2 McMahan's Demanding Understanding of Autonomy

McMahan[132] defines autonomy as "the capacity to direct one's life in accordance with values that one reflectively endorses" (McMahan 2002, 256). He agrees with Nozick that what underlies our intuition of an equal worth of all persons is the "ability to form a picture of one's whole life (or at least of significant chunks of it) and to act in terms of some overall conception of the life one wishes to lead" (Nozick 1974/2008, 50; cited in McMahan 2002, 256).

This concept of autonomy, however, may very well be too demanding and represent an example of anthropofabulation, that is, an idealization of human capabilities (Buckner 2013). It is doubtful that all persons have a picture of their lives in twenty years, let alone the whole period of time until their deaths. Moreover, it is not clear why the existence of such a 'life plan' should make *all* the difference between those subjects with a right not to be killed and those with only an interest in their continued existence. Think about the Dostoevsky-reading piano-player (person A) and the wrestling-watching bowler (person B) from Pluhar's thought experiment. McMahan wants to defend the claim that both have an equal right not to be killed. I strongly doubt that this is due to them being able to form pictures of their whole lives or them endorsing certain values, however.

Autonomy is a widely used concept in philosophy and has multiple different meanings. Generally, however, the capability of autonomy is understood to be very closely related to *self-government* (e.g., Buss 2014). This description vaguely refers to the idea that a prerequisite to act autonomously is the capacity to take a 'step back' from oneself, to get into distance. Self-government, thus, seems to imply a certain freedom from one's natural dispositions.

On a very weak reading, an animal could be said to possess the ability of autonomy, if she was able to choose between several options to act, in contrast to an organism whose behavior is completely determined by internal and external factors, such as fixed action patterns (see, e.g., Beauchamp 2005; Beauchamp and Wobber 2014). On a different reading, autonomy does not merely entail

132 The remarks in this and the subsequent chapter will largely focus on McMahan's account, but many aspects apply to Garner's position as well.

the capacity to choose among different options how to satisfy a desire, but the capability to decide whether one wants to *follow* the desire in the first place. This understanding of autonomy resembles, for instance, Dworkin's definition of autonomy as "second-order capacity to reflect critically upon one's first-order preferences and desires, and the ability either to identify with these or to change them in light of higher-order preferences and values" (Dworkin 1988, 108).[133] If McMahan's demanding definition of autonomy – emphasizing a person's alleged ability to plan out her whole life in accordance with a clear vision of the kind of life she wishes to lead –, is being replaced with this more moderate one – putting more focus on a person's ability to critically reflect upon her desires –, McMahan's equal wrongness thesis becomes more comprehensible, in my view. Thus, McMahan could argue that the killing of person B should be regarded as equally wrong as the killing of person A, because both, it is assumed, have autonomously chosen the lives they are living and this decision demands respect. One way that this respect finds expression is by acknowledging that all autonomous agents have an equal inherent worth, resulting in an equal right not to be killed.

6.3.5.3 The Gradual Nature of Autonomy

An insistence on autonomy brings several other problems about. As McMahan and Garner acknowledge, autonomy is a matter of degree. This is the reason why McMahan introduced a third, intermediate stage of moral status in the first place. This stage is said to include all those beings, who are neither completely non-autonomous, nor fully autonomous agents.

I see two major problems with this approach. The first concerns McMahan's equal wrongness thesis. If autonomy can occur in degrees, then surely some persons can also be more autonomous than others. For instance, person A

---

133 Naturally, this concept of autonomy is not uncontroversial (see, for example, DeGrazia 1996, 205f). Beauchamp's (2014) main reason for rejecting Dworkin's second-order approach is that psychological research showed that our decisions much more often are a result of implicit, unconscious cognitive processes than of second-order reasoning. Our beliefs that we have come to a certain result only after careful evaluation of the pros and cons might be pure *post-hoc* fabulation, consequently. I accept this objection insofar that a pure second-order theory of autonomy will very well turn out to be too restrictive in light of current research, because it would imply that the majority of our everyday behavior is non-autonomous (Beauchamp 2005, 311; for a proposal of how to modify Dworkin's definition of autonomy, see DeGrazia 1996, 205f). Yet, it should be uncontroversial that at least sometimes we engage in reflective thinking and evaluation. Our general capability for reflective thinking is not diminished by the fact that in some (or even many) cases we err on the causal role of these reflective thoughts. To count as an autonomous agent, one does not have to act autonomously in every single situation.

from the example above might only play the piano and go to the opera, because of a involuntarily need to follow certain conventions rather than because it is in his actual will to do so. If it is assumed that person A is not as autonomous as person B, however, then it is not immediately apparent why the killing of A should not also be regarded as less bad than the killing of B in McMahan's account. That both A and B have the *ability* to act autonomous should not be sufficient to also ascribe them with an equal right not to be killed.

The second problem concerns the introduction of a third moral status. McMahan's postulation of an intermediate stage for those beings in possession of some of the rudiments of autonomy might be better able to capture the gradual nature of the ability of autonomy, but it still does not answer what exactly he argues these criteria of (rudimentary) autonomy to be. Instead, the problem of providing a clear-cut definition of what kind of abilities should unequivocally separate autonomous from non-autonomous beings is merely being shifted to the question how rudimentarily autonomous beings should be defined in contrast to fully autonomous and non-autonomous beings. The killing of these beings, that is, infants and maybe also great apes, according to McMahan (2002, 265), should neither be determined through a rights account, nor through a utilitarian time-relative interest account of killing. Rather, McMahan suggests that it will require the creation of a third normative theory.

The postulation of three different normative theories, of course, results in a very complex overall theory of the ethics of killing. In fact, it could be argued that McMahan's theory would tremendously profit from abandoning three different stages of moral status and instead argue either in a utilitarian account or an interest-based rights account, such as Alasdair Cochrane's. Such a modification could avoid the unclear relative weights of interests and rights and significantly simplify McMahan's theory. These modifications would imply the abandonment of the equal wrongness thesis, which regards the killing of two persons as equally wrong, but as even McMahan accepts that we should, in a lifeboat case, sacrifice the person which is harmed less through her death than the other person, he does not seem willing to stick to the equal wrongness thesis under all circumstances, anyway. I think that the proposals would in fact better fit McMahan's own considerations.

A dismissal of the idea of three normative theories is not to be equated with a rejection of the moral relevance of autonomy, as I will attempt to show in the following paragraphs. For McMahan, the possession or lack of autonomy determines on which one of the three stages of moral status a being is to be located. If autonomy, however, is to be understood as the ability to critically reflect on one's first-order desires, then autonomy is already *included* in the prudential unity relations. An artificial separation between autonomous, almost autonomous and non-autonomous individuals might indeed be unnecessary or, worse, not justifiable.

To strengthen this hypothesis, some more background information on an understanding of autonomy in terms of self-government and self-reflection will be needed. To do so, I will mainly refer to an essay by Neo-Kantian philosopher Christine Korsgaard, because it will enable me to link McMahan's ideas and considerations on (rudimentary) autonomy to the conceptual differentiations that have been made throughout the main part of this thesis afterwards.

Korsgaard argues that certain animals can very well act on purposes, that is, they are aware of their desires and can think about ways to pursue them. However, an animal does not "choose to pursue those purposes", rather, these are given to him "by his affective states: his emotions and his instinctual or learned desires" (Korsgaard 2006, 110). As a result, an animal is also not able to *choose* between two desires. Rather, his choice is *determined* by his affective states. In his regard, the animal is unfree.

We as autonomous agents, in contrast, choose not only the means but also the ends: "Our purposes may be suggested to us by our desires and emotions, but they are not determined for us by our affective states, for if we had judged it wrong to pursue them, we could have laid them aside" (ibid.). By doing so, autonomous agents often refer to self-interested reasons, that is, they think about what is in their best or long-term interests. To determine what is in your own best interests, in turn, "requires a kind of grip on the future and an ability to calculate that do not seem available to a nonhuman animal. Just as importantly, *acting* for the sake of your best interests requires the capacity to be *motivated* by the abstract conception of your overall or long-term good" (ibid. 102).

Autonomy, in this sense, requires being able to distance oneself from one's present desires, to ask oneself if one should pursue these desires in light of overall self-interested reasons and to act in accordance with one's decision. Animals, in contrast, cannot be motivated by a self-interest in their long-term good, Korsgaard argues. Just like Frankfurtian wantons, "they act on the instinct or desire or emotion that comes uppermost" (ibid.; see also Frankfurt 1971). Below, Korsgaard's considerations will be put in relation to the major conceptual differentiations of chapter 4.

Mental space travel can motivate a subject to sacrifice a present desire in favor of a future one, as has been argued by Boyer (2008) and others, because it allows him to pre-experience the positive outcomes. An individual capable of mental space travel is not necessarily a person in the Frankfurtian meaning of the term, however, because he could still simply act in accordance with his currently felt strongest desire. Mental space travel brings the future into the present, so to speak, so that a future desire becomes the currently strongest desire. Put differently, a wanton, capable of mental space travel, may be able to act in accordance with a future desire without knowing that he does. He would still act in accordance with whatever his currently felt strongest desire would be.

Individuals capable of anticipatory planning, in contrast, know that their present and their future desires can be qualitatively different. The birds in Correia et al. (2007) and Cheke and Clayton (2012) kept on preferentially eating pine seeds, while simultaneously caching more dog kibbles. Thus, they seemed to be well aware that they could act in a way to secure the satisfaction of certain desires they currently did not feel. Put differently, the birds were seemingly simultaneously aware of two contrasting desires, separated through time, and choose to pursue one of them.

This is an important difference to the above described subjects with mental space travel abilities, who cannot plan ahead. Individuals with anticipatory planning abilities, thus, can decide to act against their currently experienced desires and are capable of consciously rejecting a present temptation. They take certain future needs into account and acknowledge that some of these future desires can be so strong as to justify the sacrificing of certain present, but weaker, temptations. This is in stark contrast to Korsgaard's above description of animals. Still, however, these subjects' decisions are likely governed by a decision between two unequally strong desires.

Anticipating what is in one's long-term interest goes beyond the capacity to anticipate a future desire at a specific point in time. Long-term interests stretch over a certain period. Thus, an animal might be able to choose to act in accordance with a *future desire* and in contrast to a present desire, because the future desire is simply stronger, but not necessarily be able to choose to act in accordance with a *long-term interest*, which may be relatively weak on any singular point in the future, but is stronger over the whole course of time.

However, if autonomy is understood as to require, as a minimum criteria, the ability to identify with or reject certain first-order desires in light of their best or long-term interests, then the anticipation of future needs and desires might be rightfully regarded as a first step into the right direction. Put differently, anticipatory planning abilities may be considered to be sufficient as to ascribe those animals, who possess the capacity of anticipatory planning, with what could be described by McMahan as rudimentary autonomy.

Thus understood, autonomy is already included in the degree of a being's time-relative interest in continuing to live in McMahan's theory, because it is captured by a determination of her prudential unity relations. The reason for this is that to reflect on one's first-order desires and accept or reject them in light of other desires and interests involves multiple internal reference-typed connections into the future, which have a direct influence on the strength of the individual's prudential unity relations. It could be questioned, therefore, whether the postulation of three different stages of moral status on the basis of an ability already implied in an individual's time-relative interest in continuing to live is indeed necessary.

6.4 An Argument in Favor of Complexity

The previous chapters provided an overview of how utilitarian and animal rights accounts, represented by the positions of Peter Singer, Tom Regan and Alasdair Cochrane, have dealt with the permissibility of the killing of animals. Two hybrid accounts, encompassing utilitarian as well as deontological elements, were afterwards presented. Certain problematic implications of these accounts have already been laid out. In this chapter, I would like to single out one problem applying to all the theories discussed above, in my view, that is, that they tend to underestimate the challenge of a thorough ethics of killing animals. In a nutshell, I will argue that things are in fact much more complicated than many animal theorists acknowledge them to be.

My argument has two parts. The first part, presented in chapter 6.4.1, will show that common classifications that have been proposed (e.g., between persons and nonpersons) are way too simplistic with regard to the ethics of killing. All positions, that have been discussed above, try to do justice to the fact that the interest in continuing to live can manifest itself rather differently in two individuals and is thus fundamentally different from the interest in avoiding pain, for instance. Either, however, this fact is argued to be morally insignificant in normal cases (e.g., Regan) or various concepts are not sufficiently distinguished from one another, so that empirical evidence of any kind of future-orientation or intentional behavior seems to be taken as sufficient evidence for an animal being strongly psychologically connected to her future self (e.g., Singer). Alternatively, qualitative differences with regard to future-orientation are indicated but not specified and empirical studies not considered (e.g., McMahan). Rather, absence of specific cognitive capabilities is just taken as a given, so that, once again, the result is the familiar, but too simplistic classification between persons and nonpersons.

As a consequence, a dual simplification is taking place; either animal theorists tend to disclaim crucial differences with regard to future-orientation between two individuals, or they tend to downplay significant similarities. I will conclude my remarks by arguing that common classifications (e.g., between persons and nonpersons) are much too unspecific to determine the strength of an interest in continuing to live and that the focus should rather lie on a classification of prospective cognition itself, as has been attempted to do in chapter 4.

The second part of my argument, presented in chapter 6.4.2., will make the case that the argument from marginal cases (AMC), at least as it is commonly used in the debate, is not adapted for the task of providing an answer to the question under which circumstances the harm that an animal suffers through her death makes it unjustifiable to kill her for food or to experiment on her. I will argue that the AMC does not provide us with proper and thorough analysis of how much moral weight an animal's interest in continuing to live should have.

6.4.1 *The Dual Negation of Differences and Similarities*

Regan and Singer both introduce a criterion to distinguish those beings with an interest in continuing to live from those without. However, they make their criteria, on the one hand, too strong – why should *only* those animals be harmed by their deaths, who are subjects-of-a-life and persons, respectively? – and, on the other hand, too weak – why should the harm through death of one individual, who does not even know that she *has* a future, be comparable to another individual, who makes all kinds of efforts and sacrifices with the purpose to autonomously shape and organize her own future, for instance?

Whether or not a being can be harmed through her death to begin with is, according to Singer, a matter of whether or not she is self-conscious, where self-consciousness is defined with Locke as the ability to conceive of oneself as a distinct entity with a past and a future. A self-conscious being is "capable of having desires about [her] own future" (Singer 2011, 76). Thwarting these desires by killing a self-conscious being is what makes it wrong to kill her. Yet, Singer aims to include as many animals as possible into the group of persons, so that he has a considerable interest in defining the relevant kind of self-consciousness extremely minimal.

Singer implicitly seems to assume (rightly, I believe) that self-consciousness, as defined above, is a specific kind of self-awareness. Not just *any* evidence of self-awareness in a being should suffice, therefore, to show that its killing is wrong. In spite of this, Singer later seems to change his definition of persons insofar that it is no longer the (rather demanding) Lockean self-consciousness that separates persons from nonpersons, but rather any level of self-awareness or an awareness of the future. Indeed, Singer at one point explicitly argues that those animals who are self-aware sufficiently meet his (i.e., Locke's) definition of a person (Singer 2011, 103).[134]

As a consequence, when listing the behavioral evidence for at least some nonhuman animals being persons, Singer refers to several abilities, which all *relate* to the broader concept of self-awareness, but not all of which seem to be identical with or necessarily indicate the kind of Lockean self-consciousness, which technically should represent the *only* morally relevant kind of self-awareness for Singer. In detail, he mentions several empirical studies, observations and anecdotes, that concern very different abilities such as sign-language and other forms of communication, self-recognition (e.g., mirror tests), the conscience of one's own social position in comparison to other group members, a sense of time, anticipatory planning, problem solving, (sequential) tool use and others (ibid. 95–103).

It is doubtful that Singer actually holds that every single one of these very different abilities provides sufficient evidence for the relevant animal being conscious of herself as a distinct entity with a past and a future. At the very least, however, he seems to suggest that any behavioral evidence indicating some level of self-awareness in an animal, together with a willingness of giving animals the benefit of the doubt[135], is sufficient to define the relevant animal as a person. Consequently, Singer comes to the conclusion that mammals, and possibly also birds, fish as well as octopuses are persons.

To illustrate Singer's problematic argumentation here, two of his own examples will be singled out. The first example refers to the problem that multiple, very different abilities, broadly related to self-awareness, are acknowledged to provide sufficient criteria for personhood, even though Singer should technically be solely interested in the Lockean self-consciousness with its emphasis of diachronic consciousness and the ability to be aware that one has a past and a future. The second example illustrates that Singer is not even consistent in what behavioral evidence exactly he accepts as sufficient evidence for any one of these abilities, exemplified by his use of the concept of planning.

In discussing whether and which kind of animals are persons, Singer refers, among others, to the abilities to recognize oneself in the mirror (ibid. 95), to

134 In a different passage, Singer even gives the impression of equating "self-awareness" with "future-directed thinking" (Singer 2011, 101).

135 *Benefit of the doubt* here refers to the argument that if we are not entirely sure whether a certain individual is a person or not, we should rather err on the side of caution, i.e., preferably include too many beings into the class of persons, than erroneously exclude certain beings. Wrongly ascribing a being with a higher moral status than she should rationally have is argued to be less severe than wrongly denying her a higher moral status.

anticipate recurring events (e.g., Christmas or breakfast[136]) (ibid. 95f and 102), to be conscious of one's own social position in the pecking order (ibid. 102) and, finally, to the famous chimpanzee Washoe, who "put signs together to form simple sentences and, in doing so, provided strong evidence of a sense of self" (Singer 2011, 95). The abilities that seem to underlie these animals' behaviors seem to be significantly different from one another, however, and, crucially, do not all indicate the presence of a diachronic self-consciousness of some kind.

Furthermore, Singer lists several very different behaviors that allegedly all indicate animal planning capabilities, for instance, the cooperative behavior of chimpanzees working together to overcome an electric fence with the aim to reach palatable leaves in a tree (ibid. 98), the effort of an octopus carrying coconut half shells over long distances, which Singer argues to indicate "that the octopus is aware of its own future need for shelter and is planning ahead" (ibid. 103), the self-control of a pig resisting getting a hidden food until her stronger companion, who is unaware of the location of the treat, is either out of sight or in greater distance to the hidden food (ibid. 99), but also Santino's stone-throwing and the specific-satiety paradigm experiments with jays (ibid. 99f), that have been described in chapter 5.1.3 in detail.

Yet, Singer also acknowledges that a dog "anticipating going for a walk" by putting her leash at the feet of her owner does not suffice to indicate the dog's "ability to imagine a future event" – which Singer defines as mental time travel –, because the dog "is simply in the grip of her desire to go for a walk and is acting on that desire", whereas human adults "can plan on satisfying motivations that they do not presently feel – as when we go shopping to make sure there will be something to eat for dinner, even though we are not hungry now" (ibid. 98). It is, however, not self-evident whether, for instance, the chimpanzees and the octopus from the examples above, in fact differ in this regard from the dog.[137]

To sum up, Singer does not conceptually separate between all these extremely different cognitive capabilities he mentions but rather seems to acknowledge each one of them as (at least) strong indication for the relevant individual fulfilling the criterion of being a person. Moreover, his concept of

136 From the anecdotal observation that hens, who have been locked up in the stall at night, are "eager to get outside in the morning", Singer infers that they may indeed be capable of "anticipating the future" (Singer 2011, 102).

137 Singer seems to give considerable moral weight to the fact that "[m]any people who live with dogs and cats are convinced that their animal companions are self-conscious and have a sense of the future" (Singer 2011, 102), however, so that he is willing to include dogs and cats in the group of persons as well.

planning, among others, is unspecific and inconsistently used. As has become evident in chapter 4, however, even those abilities actually suggesting an ability to engage in future thinking on behalf of the individual can be extremely different from one another and should not be lumped together under one umbrella term.

Singer does acknowledge that the distinction between persons and non-persons is not categorical but a matter of degree, but he fails to work out the crucial consequences of this observation for his theory of the wrongness of killing. The gap between, on the one side, a psychologically unconnected being, completely restricted to the present moment, and, on the other side, a being with a linear conception of time, is indeed very wide.

The discrepancy of an insufficient conceptual differentiation becomes even more evident in Regan's position, as he does acknowledge that the interest in continuing to live can be weaker or stronger, but negates this fact insofar that he does not deem it to be significant, at least not under normal circumstances. Like Singer, Regan lists many abilities that are significant for how much harm an individual suffers through death, but fails to make the differences that these abilities plausibly entail fruitful for his theory. To conclude, Singer as well as Regan accept that interests in continuing to live vary, but then go on to negate this fact by either judging these differences to be insignificant (Regan) or by not further conceptually differentiating between the abilities that have an impact on this interest (Singer).

Garner and McMahan make quite a similar move, but from the other side of the gap, so to speak. Both picture animals as living completely in the present (e.g., Garner 2013, 126), both fail to explain, however, what this hypothesis really amounts to. Both authors acknowledge that animals pursue goals and may also reason about the means to reach them. Yet, both agree that animals cannot sacrifice first-order desires in light of the individual's overall conception of a good life.

This classification, that seems to be influenced by Harry Frankfurt's (1971) separation between persons and wantons, is way too simplistic, however, as has already been argued above. There are not just those beings, on the one side of the gap, who are trapped in the present, and those beings, on the other side, who are fully autonomous beings. It is time to regard animals as what they are, neither as full-blown persons, being just as much harmed through their deaths as paradigmatic human adults, nor as (replaceable) beings, 'hopping' from one isolated experience to the next, unable to form any kind of conscious psychological connection to their past or future.[138]

138 Cochrane, whose position is largely based on McMahan's time-relative interest account, gives much more credit to the wide range of cognitive capabilities of animals than

Someone might acknowledge my above considerations but still argue that from a moral point of view, the tedious undertaking of developing a detailed classification of prospective cognition is in fact unnecessary, after all, because obviously, even an extremely weak interest in continuing to live is sufficiently strong to justify a right not to be killed or, in non-rights accounts, a strong protection from being killed. Recall that Regan and Singer attempt to show that animals are *strongly* harmed through their deaths, whereas McMahan, Garner and, with reservations, Cochrane, repeatedly emphasize that animals are only very *weakly* harmed through their deaths. Interestingly, and somehow surprisingly, they all come to relatively similar results with regard to the practical implications. How can that be possible?

Virtually everyone concerned with the morality of killing animals refers to the argument from marginal cases. It seems to answer exactly the question attempted to be solved, that is, when harm an individual suffers through death becomes sufficiently strong to justify a right not to be killed or, in non-rights accounts, a strong protection from being killed. And maybe it is even unnecessary to define the *exact* magnitude of this harm for each and every individual animal and human being, if we simply focus on those human beings with an arguably only weak interest in continuing to live; for instance, severely cognitively disabled human beings[139]. If it turns out that even these human beings enjoy a robust right not to be killed, then it seems compelling that those animals with an arguably stronger interest in continuing to live should enjoy at least the same moral status than severely cognitively disabled humans. And as this status is evidently very high, animals should be ascribed with an equally strong protection from being killed.

This line of argumentation can be highly problematic in the context of killing, however, and is not always persuading, as I hope to show in the following chapter.

McMahan does. For instance, he states that "it would be a mistake to suggest that animals are entirely 'trapped in the present'" (Cochrane 2007, 299), and claims that animals differ from human persons merely with regard to the *number* of their psychological connections over time, whereas McMahan seems to think the majority of animals is unable to form internal reference-typed connections at all.

139 Of course, Regan, e.g., would argue that these beings, provided that they are subjects-of-a-life, have a right not to be killed not for the reason that they have an interest in continuing to live, but because they have exactly the same inherent value as all other subjects-of-a-life. This position has repeatedly been criticized, however, for justifying the ascription of rights on a rather mysterious basis. Thus, I will presuppose that an individual's right not to be killed is based on the (more or less strong) interest in continuing to live (Cochrane) or that the consideration of this interest imposes at least very high moral obligations on us as moral agents, respectively (Singer, Garner, McMahan).

6.4.2 *A Word of Caution Concerning the Argument from Marginal Cases*[140]

The *argument from marginal cases* (AMC, for short)[141] is one of the most influential arguments in the animal ethics literature. It has been included in a collection of one hundred most important arguments in Western philosophy (Bruce and Barbone 2011). The core purpose of the AMC is to argue against moral anthropocentrism. Moral anthropocentrism argues that it is justified to give priority to human over animal interests on the ground that there is a morally relevant criterion determining the higher moral status of those beings satisfying it, and which is possessed by *all* human beings and human beings *alone*. If it turned out that there is no such plausible criterion, then moral anthropocentrism would be an instance of speciesism, that is, an arbitrary, unjustified prioritizing of human over animal interests.

Under the presupposition that this morally relevant criterion may neither be definitional – being a member of the biological species *homo sapiens*, for example – nor metaphysical – being in God's own likeness, for instance –, then this criterion may either refer to intrinsic properties, such as certain cognitive capacities, or relations[142] these individuals stand in to one another. Consequently, a plausible defense of moral anthropocentrism either refers to certain properties or capacities human beings have, or to relations persisting between them. Moral anthropocentrism, thus, cannot be defended by recourse to definitional or untestable criteria (Horta 2014, 144).

Those in favor of the argument of marginal cases now argue that such a criterion does not exist and that, as a result, the prioritizing of human over animal

140 I am especially grateful to Markus Wild for helpful criticism and comments on this chapter.

141 The term 'argument from marginal humans' has been criticized for being offensive and also conceptually imprecise and misleading. Alternative phrases have been suggested, such as Horta's (2014) "argument from species overlap". I agree that the term is extremely problematic. As, however, *argument from marginal cases* has become a technical term, and as such, as Garner (2013, 177, fn 1) rightfully observes, enables readers to follow the debate easily, I will stick to its original name. By referring to 'marginal cases' or 'marginal human beings' I do in no way intend to question these humans' species or group membership. Generally, the group of marginal cases is said to include more or less severely mentally disabled people, permanently comatose, newborns and small infants, late-term fetuses and sometimes also anencephalics.

142 The argument from marginal cases has traditionally been viewed as being vulnerable to relational positions. Horta (2014) argues, however, that even if emotional and other bonds are included in the definition of a criterion, the problem remains that there always will be human beings to whom we do not have such a relation or who cannot establish such a relationship themselves. He concludes, thus, that the argument from marginal cases also applies to relational positions.

interests cannot be justifiably defended.[143] Since Evelyn Pluhar's (1995) analysis, two versions of the argument are typically discriminated: the biconditional or weak version of the argument states that *iff* marginal human beings have a high moral status, then those animals, who are morally relevantly similar, have an equally high moral status.[144] This naturally allows for the theoretical possibility of lowering the moral status of marginal human beings. A somewhat stronger version, that is, the categorical version of the argument, argues that *because* marginal human beings have a high moral status, those animals, who are morally relevantly similar, have an equally high moral status.

The argument from marginal cases has often been used in this rather abstract way, questioning the anthropocentric view that human interests *per se* have a higher moral standing than animal interests, just because they are human (see, above all, Singer 2009, 2011). It has also been applied to a more specific context, however. With regard to the infliction of pain, the argument is particularly compelling. There are good reasons to assume that all sentient beings have a more or less equal interest in avoiding pain.[145] If we accept that the morally relevant criterion for why it is generally wrong to inflict pain on others is their sentience, then we are not justified in conducting painful experiments on animals but not on mentally disabled human beings.

In a similar fashion, the argument from marginal cases can be applied to the morality of killing animals. However, its application to the context of killing may be argued to be more challenging than in terms of the infliction of pain. Singer, for instance, professes:

---

143 Naturally, opponents of the argument from marginal cases strongly object to this premise. The debate around the AMC has become much too complex as to do it justice here. Luckily, it is not relevant for my argument. It will be enough to accept the argument's premise and assume that there is indeed no such a morally relevant criterion that applies to all and only human beings.

144 (*High*) *moral status* here is used as a more neutral phrase to avoid terms such as rights, which would be rejected by utilitarians.

145 Many factors can contribute to or dampen the experience of pain and suffering. It is generally accepted, however, that humans do neither *per se* suffer more from pain or distress than animals do, nor that persons suffer more than nonpersons. Even though there certainly are causes for suffering or distress only accessible to (human) persons, there are also circumstances under which some animals suffer much more than other animals or human beings. It is plausible to assume, for instance, that a prey animal, for which immobility represents another substantial stress factor, suffers considerably more from a broken limb than an adult human (presumed that she of all people is not a professional marathon runner), who additionally has the capability to mentally focus on brighter days in the future or who knows that in a few weeks, the pain in her leg will decline.

> The application of the principle of equality to the infliction of suffering is, in theory at least, fairly straightforward. Pain and suffering are in themselves bad and should be prevented or minimized irrespective of the race, sex, or species of the being that suffers. How bad a pain is depends on how intense it is and how long it lasts, but pains of the same intensity and duration are equally bad, whether felt by humans or animals. The wrongness of killing a being is more complicated. (Singer 2009, 17)

This is so for two main reasons, in my opinion. First of all, investigations of the wrongness of killing are more challenging than the wrongness of the infliction of pain, to begin with. The relevant criterion(s) for what makes it wrong to kill a being has been vigorously debated and there may be few, if any, unanimously acknowledged intuitions here. Whereas, for example, most people would agree that death can do more harm to some beings than to others, not everyone takes this to be morally relevant, at least not in normal cases. Similarly, autonomy is sometimes, but not universally, argued to represent a further crucial criterion. That their sentience makes two otherwise cognitively very different individuals morally relevantly similar regarding the infliction of pain, in contrast, is a widely shared belief and has great persuasive power. As a result, there will be many different theories about what exactly makes two individuals morally relevantly similar with regard to their deaths.

From this it does not follow, of course, that the AMC can never be applied to the wrongness of killing, but it will have much less persuasive power than it has in the context of the wrongness of pain. As the AMC gains its particular forcefulness from uncovering an inconsistency in our thinking and convincing us to give up certain beliefs in favor of other stronger, commonly shared beliefs, this is an unwelcome consequence. What makes the AMC so compelling with regard to the equal consideration of pain is that it is absolutely plausible to assume that it is sentience, first and foremost, that should be morally relevant here. Even though some may still have the intuition that human pain 'somehow' matters more than animal pain, this view will be particularly difficult to sustain in the face of the claim that pain matters, regardless of who experiences it. Here, the AMC can reveal certain underlying irrational and unjustified beliefs and thus uncover them as exemplifications of speciesism. What exactly makes killing wrong is much more disputed, however, so that the AMC runs the risk of losing much of its persuasive power.

Second of all, it should be kept in mind that the group of marginal cases may not be homogenous with regard to its moral relevance (see also Regan 1982, 143f and Garner 2013, 148). Most infants, for instance, will develop the cognitive capacities that are commonly argued to be definitional for persons and it is open to dispute wether the potentiality of personhood should be taken into account. Also, two adults with trisomy 21 can tremendously differ in terms of their cognitive capabilities from one another, due to the four subforms of the syndrome, and they certainly differ hugely from permanently comatose or severely cognitively disabled adults. The same applies to the broad group of animals, of course. While these differences are much less relevant with regard to the infliction of pain, however, because sentience is a shared feature by most or all of these otherwise quite different individuals (with the exception of, for instance, anencephalics), they can play an important role with regard to the wrongness of killing. This means that it can be notoriously difficult to compare the harm of death for one marginal case with another one, let alone with an animal's.

Even though this problem can, in theory, be circumvented by identifying subclasses of marginal cases and animals, of course, the heterogeneity of the group of marginal cases poses another challenge to someone wishing to apply the AMC onto the context of the wrongness of killing. Again, I want to stress that this is not to argue against the logical coherence of such an application.

A non-interest based rights account like Regan's can avoid the above mentioned problems to a certain extent, because it defends an equal right not to be killed that is independent from the respective harms the individuals suffer through death. At the same time, Regan must rely on the forcefulness of the argument itself to a much greater degree, however, because the *equal* inherent value of all subjects-of-a-life, regardless of their differing psychological architectures and also regardless of the strength of their interests in continuing to live, can only be expanded onto animals, if its validity with regard to human marginal cases is accepted. Regan, therefore, must *necessarily* defend the strong, categorical version of the argument. Due to the heterogeneity of the marginal cases group mentioned above and due to the actual ambiguity of our intuitions here (to be exemplified below), not everyone would be willing to accept the premise that all human beings have an equally high moral right not to be killed, however.

Having sketched some problems surrounding the argument from marginal cases in the context of the wrongness of killing, let us have a closer look on one specific argumentation against lethal animal experiments by Alasdair Cochrane. Cochrane presents this argument in the third chapter of his main work *Animal Rights Without Liberation* (esp. on pp. 68f) and in an earlier essay with the title "Animal rights and animal experiments: an interest-based approach" from 2007.

In discussing lethal therapeutic experimentation, Cochrane argues, in accordance with his interest-based rights account, that the permissibility in each case depends on whether the benefits of a particular experiment are sufficiently strong to outcompete the specific animal's interest in continuing to live. Crucially, thus, Cochrane acknowledges that there can be certain benefits, at least in theory, that would justify a non-consideration of the animal's interest.

As Cochrane already established only a weak interest in continuing to live in most animals, it might be expected that the great outcome of such an experiment indeed overrides the animal's interest (Cochrane 2012, 68). But if it is legitimate to override an animal's weak interest in continuing to live, the argument from marginal cases admonishes, then it must also be legitimate to do the same with humans with an equally weak interest in continuing to live. There *are* human beings with an equally weak interest in continuing to live, Cochrane goes on, for instance, babies and severely mentally disabled human beings. Therefore, if it is permissible to use animals for lethal therapeutic experiments, then it is also permissible to use infants and seriously cognitively disabled human beings for these experiments. As most of us share the belief that the conduction of painless but lethal experiments on these humans is "too costly and horrific to embrace" (Cochrane 2007, 310), we should conclude that the (weak) interest in continuing to live of both human marginal cases and animals is sufficiently strong as to establish a right not to be killed by experimentation (Cochrane 2012, 69).[146]

Even if the problem of thwarted potentiality in case of infants is disregarded and the focus solely lies on cognitively disabled human beings, I believe that this line of argumentation is highly questionable. Cochrane considers the fact that we, as a society, do not allow experimentation on human marginal cases to be a direct consequence of their degree of interest in continuing to live. Thus, he concludes that even though this interest is admittedly relatively weak, it allegedly is sufficiently strong to establish a right not to be killed by experimentation, *because* we refuse to use human marginal cases. A weak interest justifies a strong right, thus.

I will argue that this move requires further justification, however, and that an appeal to intuition is not sufficient here. How two morally relevantly similar individuals are to be treated in a specific situation and whether they may be used for lethal experiments is a matter of their moral status. This moral

146 Compare this also to McMahan's (2005) argument against animal experimentation, summarized in chapter 6.3.4. In proposing that an experimenter should ask herself whether she would conduct the same experiment that she conducts on animals on a severely retarded infant, orphaned at birth, McMahan transforms the utilitarian principle of the aggregation of interests into a subjective moral conflict. Indirectly, however, he concludes that the experimenter's refusal to kill the child evidently showed that even a weak interest in continuing to live was sufficiently strong as to ground a very strong obligation in us.

status, in turn, is based on the moral theory at hand. Whether one should lower the moral status of marginal cases, raise the moral status of animals or define a new moral status for both marginal cases and animals, thus, is not a matter of intuition or personal preference, but should be answered by the underlying theory. A moral theory should then either acknowledge its practical consequences for animal (and human) experimentation or alternatively make adjustments. The following paragraphs will attempt to clarify this claim.

In each moral theory of the wrongness of killing, there is a moral criterion (C) which lies at the heart of why it is wrong to kill an individual; for instance, because it violates her interest in continuing to live or her right to respectful treatment. As a shorthand, one might say: it is *prima facie* wrong to kill a being fulfilling C (e.g., because she has an interest in continuing to live). Based on this criterion, the moral status (MS) regarding the life of the individual is to be assessed. Moral status, in this context, is to be understood as providing moral rules concerning the (im)permissibility of killing. The moral status, for example, identifies the circumstances or exceptions under which the individual may be killed. MS states, thus, that it is permissible to kill a being fulfilling C, iff y. How we are allowed to treat an individual (T) in a specific situation, for instance, whether we are allowed to use it for a lethal experiment, can be inferred from her moral status, therefore. T states, that it is (not) permissible to kill a being with x, because it is (not) the case that y.

The argument from marginal cases, then, notes an inconsistency: two beings who, according to the criterion, should have the same moral status, are in fact being treated unequally. A rat, for example, is routinely killed in experiments, whereas a severely cognitively disabled human infant, that is, a marginal case, is protected by a right not to be killed that is as strong as it is for *any* other human being.

There are several possibilities to solve this inconsistency. One could argue, for example, that the two different treatments are in place and that the criterion (and, accordingly, the determination of the equal moral status) must be wrong. Consequently, the theory is being adjusted and a criterion is found which allows the attribution of a high moral status to the infant but not to the animal. As any proponent of the argument from marginal cases argues against moral anthropocentrism, this possibility can be disregarded. Rather, it will be assumed that the criterion is correct and the two individuals in fact have an equal moral status. Thus, treating them unequally must be wrong.[147]

147 McMahan proposes to regard our favoring of disabled human beings over animals of similar mental capacities as acts of supererogation, because "we are morally permitted to lavish resources on the care and comfort of the severely retarded, even if we decline to

The question then is how the two beings should be treated instead. If moral status is understood in the sense described above, that is, as identifying the circumstances and exceptions that justify the killing of an individual with an interest (or right) not to be killed, then one must adjust the two individuals' moral status.

At first glance, it seems that in doing so three different proceedings are possible. One could

(i) lower the moral status of marginal cases
(ii) raise the moral status of animals, or
(iii) defend a convergent assimilationist position[148] implying a new middle moral status for animals as well as marginal cases.

Proponents of the biconditional version regard all three as logical coherent solutions. Yet, the decision for one of the three (typically the second one) is commonly represented to be one of *intuition*. The prospect of using humans in lethal experiments is so terrifying, Cochrane argues, that we should rather raise the moral status of animals than to allow for human experimentation. This implies that the correct moral status of animals can be directly inferred from an observation of how marginal cases are *de facto* being treated in today's (Western) society. I will argue that this is extremely problematic, because it solely rests on the assumption that the fact that marginal cases are not used for lethal experiments can be exclusively deduced from their high moral status regarding their lives.

This assumption is questionable for one thing, intuitions concerning the moral status of marginal cases vary, as will be exemplarily shown below by help of late-term abortions. Furthermore, it is quite possible that even widespread intuitions concerning the prohibition of killing human marginal cases do not exclusively rest on our specific moral theory of the wrongness of killing, but

---

do the same for animals with comparable capacities" (McMahan 2002, 231). As McMahan himself admits, we need to be extremely careful, however, "that our partiality for the severely retarded does not have the effect of dulling our sensitivity to the status of animals" (ibid.). He thus acknowledges the relevance of special relationships, but only insofar that they can prompt us to go beyond what we are morally obliged to do. To regard our special care for human marginal cases as supererogatory acts has problematic implications and invites several objections. First and foremost, it is not clear whether McMahan's proposal really avoids the error of speciesism or if it once again unsuccessfully attempts to rationalize our arbitrary favoring of one group of beings over another one. Furthermore, the proposal begs the question how it can be made sure that the favoring of human marginal cases does not come at the expense of their animal counterparts, i.e., that the animals are not being put at a disadvantage.

148 This term is used by, e.g., McMahan (2002) and Cochrane (2012). Singer (2009, 2011) argues for a similar position.

on additional, highly emotional reasons, which might make a rational consideration of these marginal cases' moral status difficult, if not impossible. I will precede by further clarifying these two statements.

Our intuitions with regard to the moral status of marginal cases are far from being unambiguous, which is particularly, but not exclusively, problematic to defenses of the categorical version of the argument from marginal cases. People tend to have very different beliefs about, for instance, the permissibility of late-term abortions, infanticide[149] or euthanasia. Religious beliefs can explain some of these variations, but certainly do not fully account for them. An atheist may very well defend an unborn infant's right not to be killed that is as strong as it is for the mother, making late-term abortions[150] permissible only in extremely rare cases, in which the life of the mother would be in concrete danger, whereas a Christian might hold that the mother's right to self-determination should have more moral weight than the unborn's right and that only the mother has the authority to choose whether she will give birth to her infant or not. Furthermore, and importantly, even one single individual may have *contrasting* intuitions on this matter, depending on the specific situation and context, resulting in a feeling of undecidedness and ambivalence rather than in strongly held convictions and opinions.

This ambivalence regarding the permissibility of (late-term) abortions is also apparent in jurisdiction. Even if one takes into account only Western European countries, laws show a huge diversity (Budde 2015, 12, Tab. 3.1). To complicate things further, *de jure* and *de facto* regulations often widely diverge

149 Admittedly, intuitions regarding the wrongness of killing healthy, non-disabled infants are very stable. Infants, however, raise the problem of potentiality, because even though they currently do not possess certain higher cognitive capabilities such as autonomy, rationality, future thinking etc., they will very likely develop them sooner or later. Even if two people might agree on infants possessing a high moral status regarding their lives, they might do so for different reasons, therefore. This does not affect the plausibility of the AMC within one concrete theory, because the AMC only states that two morally relevantly similar individuals possess the same moral status and does not make any claims about the specific criterion or criteria, that make(s) them morally relevantly similar. It does affect, however, which kind of animals, if any, should be regarded as being morally relevantly similar to (potentially autonomous) infants. Someone arguing for the moral relevance of potentiality might be skeptical whether there are any morally relevantly similar animal counterparts *at all*. If, in contrast, their potentiality is argued to be morally irrelevant, then many more animals will probably be allowed to have an equal moral status than the infants, depending, of course, on the criterion. Again, this strengthens my point, thus, that intuitions concerning marginal cases can actually vary to a great extent.

150 Late-term abortions or post-viability abortions describe the termination of the pregnancy after the twenty-second week of gestation. After 22 weeks, the fetus is potentially able to survive outside the womb (Budde 2015, 21).

within one country; a state may have a strict law, but a deliberate practice, as is the case in Great Britain, or *vice versa*, as, for example, in Austria. Spain has quickly moved from a total prohibition of abortions, that made a termination of a pregnancy not even permissible in cases, in which a continuous pregnancy would endanger the mother's life, to a very liberal law (ibid. 9f).

The German law concerning late-term abortions can be interpreted as to exemplify the attempt to accommodate various contrasting beliefs and intuitions on the matter.[151] On the one hand, it represents the wish to ascribe the unborn infant with a very strong right not to be killed, independently of her mental abilities and disabilities. On the other hand, the medical implications, which make a late-term abortion possible at any point of time during the pregnancy, if the woman's physical or mental health is at risk, are defined very weakly in practice. In fact, the abortion of a (potentially) cognitively disabled fetus, for instance, can be conducted right until her birth on the grounds of the risk of psychological distress that the infant's (potential) disability might have on the mother's mental health (ibid. 28f). Put differently, it is very doubtful that any woman would *de facto* be *forced* to bear her child against her will, even though the *de jure* regulation makes late-term abortions possible only in exceptional cases.

The woman's right to self-determination, it could be argued, outweighs the fetus' right not to be killed, thus. Emma Budde concludes that the internal ambivalence of the German law represents the impossibility to unify several diametrically opposed moral beliefs and can be read as the attempt of a compromise between irreconcilable positions (ibid. 32).

Naturally, laws or descriptive observations are not to be confused with moral beliefs. However, they sometimes serve to indicate that in some cases there are no universally accepted intuitions available and that, in contrast, people might even have several contrasting intuitions on one and the same matter, resulting in a feeling of uncertainty rather than a strongly held moral intuition. The wrongness of killing marginal cases seems to be one such a case. Some people will be more willing than others to accept that not all human beings may have an equal right not to be killed, for example, and, consequently, be much less reluctant to lower the moral status of marginal cases than others with a firm belief in the equal moral status of *all* human beings. As mentioned above, there is no consensus concerning the criterion that makes the killing of

151 The German law generally penalizes any abortion, but defines several conditions, under which an abortion will remain unpunished. The full particulars of this legal regulation are complicated but it is generally agreed that even though an abortion formally is not permissible, it can be regarded *as if it was* (Budde 2015, 15).

a being wrong, nor about whether there may be *multiple* morally relevant criteria. In sum, it is unclear how to convince one side from the need to change its beliefs, if intuitions are given the power of decision; more so, if these intuitions are far from being as clear-cut as one would want them to be.

Finally, the possibility should be considered that our intuitions concerning the killing of human marginal cases are not, or at least not exclusively, a consequence of the criterion our theory rests on, but might additionally be based on a variety of other reasons, which we, importantly, may even judge to be *morally irrelevant* or only *partly* relevant. In fact, experimentation on humans is embedded into a rich historical and cultural background and this background might (at least partly) explain why we today reject experimentation on humans. If this is correct, however, then the mere fact that we reject using cognitively disabled human beings for lethal experiments does not automatically allow the general inference that obviously even a weak interest in continuing to live justifies a strong right not to be killed. This claim will be further clarified in the following.

Many states besides Germany share an inglorious heritage of experimentation on human beings, who did or could not give their consent, either because they did not yet understand they were experimented on – as in the many reported experiments on (institutionalized) children –, they did not know that they took part in an experiment – as happened in the infamous Tuskegee syphilis cases in Alabama between 1932 and 1972 –, or they were simply robbed of their right of self-determination – as, for example, has repeatedly been the case with prison inmates in several countries.

In discussing lethal experimentation on humans, I believe that we inevitably – voluntarily[152] or involuntarily – argue against this emotionally laden background. Our refusal to use marginal cases for experimentation is probably only partly owed to considerations on the strength of their interest in continuing to live, even if, theoretically, this interest may be regarded as the decisive criterion for the wrongness of killing non-autonomous beings, as, for example, McMahan argues. Thus, the fact that we reject experimentation on humans does not automatically tell us anything about the moral significance of their interests in continuing to live, let alone about the moral status of animals.

152 In the 1970s, the German Federal Constitutional Court (*Bundesverfassungsgericht*) explicitly referred to the national historical context in their justification of prohibiting late-term abortions. In their verdict, the judges expressed their ambition to set the new German law apart from the nazi ideology that there can be lives 'unworthy' of living (Budde 2015, 17).

If this is correct, however, the topic of human experimentation is rather unsuitable to provide an answer to the question how strong an interest in continuing to live must be to establish a high moral status regarding an individual's life, contra Cochrane. Instead, I will argue that the claim that a *weak* interest in continuing to live results in an *equally strong* protection from being killed as it is possessed by subjects with a much stronger interest in continuing to live, requires further justification and might result in the need to modify one's underlying theory of the wrongness of killing.

Under the assumption that whether an individual may permissibly be killed in an experiment is a matter of her moral status, and also assuming for the moment that this moral status is a matter of degree in that some beings can have a higher moral status than others, there is a sense in which one particular moral status should be in logical accordance with the circumstances under which the killing of a specific subject should be permissible or not. Assume that our theory leads to the conclusion that the criterion for what makes killing wrong was, first and foremost, the strength of the relevant being's interest in continuing to live and also allows, in principle, that this interest can be overridden by competing interests and considerations. Then it would be up to justification, to say the least, to ascribe a certain individual with an only very weak interest in continuing to live (e.g., an animal or a marginal human case) with a strong protection from being killed *nonetheless*; a protection that would in fact be *as strong* as it were for other beings with a much *stronger* interest in continuing to live.

This is not to say that such a strong protection would be impossible to defend, of course, for one could establish a threshold which made the killing of all beings above it equally wrong; but such a move required further justification and not just appeals to intuitions. In fact, Cochrane – in contrast to McMahan – is indeed able to plausibly defend an abolition of lethal animal experiments and to explain why beings with an, at best, modest interest in continuing to live should be treated similar to those with a very strong interest.

According to Cochrane, it is *prima facie* wrong to kill a being with an interest in continuing to live, because this interest grants her a right not to be killed. It is only permissible to kill such a being, iff there are compelling reasons to do so. Because a weak interest in continuing to live only gives rise to a weak right not to be killed, however, he argues that this right can be outweighed by competing considerations. If the relevant right is only weak, relatively moderate competing considerations should, in principle, suffice to justify a violation against this right.

Yet, Cochrane also holds that we are only allowed to override an individual's right not to be killed in situations of self-defense. As has become clear in his discussion of the rodent invasion in chapter 6.2.5.3, Cochrane defines these situations of self-defense relatively strictly in that we are obligated to search for alternatives. Only if no other option is available, the violation of an individual's right not to be killed is permitted. In line with this considerably strong claim, the question whether the usage of animals in lethal experimentation should be permitted or not should not be determined by whether there are any other interests and considerations that might outweigh the animal's interest in continuing to live, contra to what Cochrane himself sometimes argues. The relevant question should rather be if the specific experiment can be understood as a situation of self-defense and whether there are alternatives to using and killing the animal. Instead of referring to the argument from marginal cases, it would be a better strategy for Cochrane, thus, to clarify how exactly these situations of self-defense should be understood in his approach.

Cochrane also defends a gradual right not to be killed, however. This seems to entail that that in case of individuals with a particularly strong right not to be killed, only extreme and very exceptional cases can be defined as situations of self-defense, whereas in case of individuals with an only moderate right not to be killed, these situations of self-defense are defined in a broader sense. This being said, it should be self-evident that most lethal experiments, that are currently routinely conducted, do not suffice the definition of a self-defense, even if self-defense is to be defined very weakly. The moral wrongness of lethal animal (and marginal cases) experimentation can be inferred from Cochrane's theory itself, therefore, without the need to refer to the argument from marginal cases.

McMahan cannot apply a similar kind of argumentation to his own approach, however, because he does not ascribe non-autonomous beings with a right not to be killed. If it is assumed that most animals as well as marginal human cases represent non-autonomous beings, then their interest in continuing to live can thus always be overridden by competing interests. Then, however, there is absolutely no reason why weak interests should not be allowed to outcompete a similarly weak interest in continuing to live. From this I conclude that McMahan should either accept the implication his own moral theory has, or adjust his theory accordingly by, for instance, including non-autonomous beings into the realm of moral right-holders, or by establishing a qualitative difference between the interest in continuing to live and all other kinds of interests, which made the former 'weightier' than the latter.

The above considerations should not be misunderstood as making the case that the argument from marginal cases is without use in the context of the wrongness of killing. It should not be misused as a 'short-cut', however, that is, as a favorable occasion to simply circumvent the unwelcome consequences one's moral theory might have. One important function of the AMC is that it alludes to certain problematic implications, which might motivate additional modifications of the theory of the wrongness of killing at hand. If, for instance, McMahan considers it wrong that a weak interest in continuing to live can be overridden so easily, this should give him a reason to investigate whether his theory could be rationally modified in a way that would accommodate this intuition, without leading to other, (even more) unintuitive consequences. Intuitions can motivate an author to modify his theory, if certain of its practical implications are too much in conflict with certain widespread beliefs, but they cannot serve as a justification to simply accept this discontinuity between one's theory and its practical implications.

Above, however, it was also argued that intuitions concerning the permissibility of killing marginal cases considerably vary and might even be based on considerations that are commonly regarded to be irrelevant or only partly relevant for the assessment of an individual's moral status, or considerations that might not be transferable onto animals. It is extremely plausible, thus, that no account of the wrongness of killing will satisfy each and every one's intuitions and beliefs on this matter. Still, intuitions should not be given the power to simply override a theory's consequences and implications, less so, if they turn out not to be universally shared and rather unstable at second glance.

In the remainder of this chapter, I will consider several possible objections to the above considerations and do my best to directly reply to these.

In fact, someone defending a moral anthropocentric position, may conclude that the cultural and historical reasons mentioned above indeed justified the higher moral status of human marginal cases over animals, because they provided additional reasons against using cognitively disabled humans for lethal experiments, which are not transferrable onto animals.

As a reply to this objection it should be kept in mind that there are good reasons for not determining an individuals's moral status by those kinds of considerations that are independent of the capabilities and characteristics of the individual herself. Otherwise we risk arbitrarily drawing a line between two groups of beings, solely based on prejudices or existing power structures. This is at the core of what has been characterized above as moral individualism and which has already been established as a precondition in the beginning of chapter 6. Instead, cultural and historical considerations can provide additional indirect reasons against the killing of a being, just as third-party interests or special relations might be acknowledged to do, but they should leave considerations about the moral status of a being untouched.

Alternatively, it could be objected that even if our intuitions concerning members of our own species may not always be justified, we still regard them as so fundamental as to not be willing to give them up. This led Garner to conclude that the AMC does not hold the reality check. He acknowledges that the argument is convincing in theory, but will never find the necessary acceptance in society; we will simply not *allow* that some members of our species be regarded as morally inferior to other members. And if we will not do so, we should not make it a part of our theory (Garner 2015, 220; Garner 2013, 152).

Certainly, the overwhelming majority of people agree that permitting painless but lethal experiments on marginal cases is not an option. Yet, they might not necessarily do so for the same reasons. Slippery slope arguments (Where should the line be drawn between those beings who can be experimented on and those who should be ascribed with a right not to be killed?), the consideration of third party interests (How will a legalization of human experimentation affect these human beings' relatives, but also researchers, nurses and other cognitively disabled humans?) and lots of other considerations can all reject human experimentation, without necessarily arguing for an *equal* moral status of all human beings regarding their lives. Thus, Garner's claim that we strongly believe that all human beings have an equal moral status and an equal right not to be killed might be questioned.

Moreover, even under the assumption that Garner is correct in that all of us unanimously refuse to consider some human beings as morally inferior to others, it should be kept in mind that historically speaking, this is a relatively novel belief. As has been argued above, it may be based, among other things, on the fear that certain practices might lead to immoral acts in the long run, that is, on slippery slope argumentations. In categorically rejecting experimentation on a human being, regardless of her being a paradigmatic adult, a cognitively severely disabled newborn infant or an anencephalic, people might thus express their wish, among other things, to distance themselves as a society from certain clearly morally wrong practices of past societies, such as the horrific nazi experiments on prisoners and mentally disabled infants. If it is acknowledged that the moral status of an individual is not to be determined by considerations of this sort, however, then the mere fact that people find the idea of gradual moral statuses for human beings or an equally high moral status for a human and an animal individual repulsive, should be irrelevant.

Instead, it was argued above that one function of the AMC is to point out certain unintuitive or unwelcome consequences a theory has and thus serving as a motivator to modify the theory at hand in such a way that it better matches these intuitions. Such a modification might then very well lead one to ascribe marginal cases with a higher moral status than the original position did. If there are other, morally relevantly similar animals, then their moral status will be enhanced as well, however, with the possible result that using these animals for lethal experiments must be considered wrong, just as using mentally disabled human beings is.

Another objection could be that even though our intuitions concerning the proper treatment of human marginal cases may not be in accordance with one's own theory in that they should actually have a much lower moral status than we are prepared to provide them with, this inconsistency should just be accepted. Then, however, animals, who fulfill the criterion lying at the heart of the theory, should be ascribed with an equally strong moral status as marginal cases. If we simply cannot help but ascribe the members of our own species with a high moral status, which they, based on our theory's rational criterion, should actually not be ascribed with, fairness and consistency demand to equally raise the moral status of their animal counterparts. Put differently, an animal's interest should just be *regarded as if* it was sufficient. This objection could be described as giving animals the *benefit of irrationality.* Consistency trumps rationality, provocatively speaking.

I do not find this solution very convincing. To say the least, it required a lot of justification and I am skeptical of whether it could be provided. There may always be additional reasons against the killing of a particular individual (e.g., a marginal case), but these reasons must not be transferred onto another individual (e.g., an animal), if they do not apply. Assume that the cognitively disabled adult's relatives would object to his killing or that there are cultural and historical reasons for our refusal to conduct deathly experiments on all kinds of human beings. This background cannot be simply transferred onto the animal with the justification that fairness obliges to do so, nor, importantly, is it necessarily telling with regard to the animal's or the human being's *moral status*. If a theory's consequences are considered to be unacceptable and not in accordance with certain beliefs one wants to keep at any rate, it is a much better option to adjust one's theory than to shrug one's shoulders.

It should always be presumed that a proponent of a given theory of the wrongness of killing is convinced of his approach. If she had come to the conclusion that the implications of her theory's criterion are unacceptable, she should have changed her criterion, for instance, or otherwise adjusted her theory. If, however, she sticks to the theory, then its implications should not be abandoned lightly, more so, if underlying intuitions are ambiguous.

6.5 Intermediate Conclusion

From the above considerations, the following conclusions can be drawn. Common classifications in animal ethics, for instance, between persons and nonpersons or sentient and non-sentient beings, are not fine-grained enough with regard to the wrongness of killing. This became particularly evident in Singer's discussion of self-consciousness, in which several very different capabilities, that should be kept separated, were equated with one another. Also, the transition from how much harm death does to an animal to her actual right not to be killed possibly is much more complex than the argument from marginal cases suggests. I propose, therefore, to once more separate questions concerning the harm of death and the wrongness of killing. Instead of cutting analysis short by help of the argument from marginal cases, we should first aim to clarify how much, *really*, animals lose through their deaths before determining the circumstances under which it might be permissible to kill them, rather than simply regarding them *as if* they are strongly harmed through their deaths.

An interest-based rights account like Cochrane's has several advantages over other theories. For one thing, it gives justice to the fact that not all sentient beings are equally harmed through their deaths (which even Regan acknowledges, though he does not include this fact in his theory) and, thus, does not negate existing cognitive differences. Cochrane also explicitly refers to McMahan's TRIA, which has been argued to represent one of the most thoroughly reasoned moral theories of killing. Furthermore, he grounds the right not to be killed on a being's underlying interest in continuing to live rather than on a mysterious inherent value. Lastly, Cochrane's account rejects the utilitarian principle of aggregation, even though it is sometimes unclear, what exactly a weak right not to be killed obliges from other right holders in concrete situations, as has been discussed in chapter 6.2.5.3.[153] An interest-based rights account should be careful, thus, not to dissolve into purely utilitarian considerations.

In contrast to the hybrid accounts by Garner and McMahan, the capacity of autonomy is acknowledged to be an implicit part of Cochrane's theory itself, because autonomy affects the strength of the prudential unity relations a subject has to her future self, as has been argued in chapter 6.3.5. Cochrane's theory has the major advantage, therefore, that it does not need to assume two or even three moral theories for the group of beings that can be harmed through their deaths and which do not fundamentally differ from one another.

---

153 Of course, the problem of an exact determination of the practical implications of a moral theory applies to all other theories besides Cochrane's as well (with the possible exception of abolitionist accounts like Regan's, where the practical implications commonly are very straight-forward).

Still, however, Cochrane uses the argument from marginal cases as a justification of the ascription of equally strong rights to beings that significantly differ from one another concerning their interests in continuing to live. As has been argued above, this move requires further justification, however. That marginal cases are not used for lethal experimentation does not automatically allow the inference that obviously their weak interest in continuing to live is sufficiently strong as to establish a right not to be killed that is as strong as it is for persons. Nor does it allow any inferences concerning the moral status of animals regarding their lives, in my view. In fact, the matter is much more complex, in my opinion, than many proponents of the AMC argue it to be. If my considerations have been correct, then the AMC can function to reveal a discontinuity in Cochrane's theory of the wrongness of killing, however, which calls for a clarification of the situations of self-defense, in which an animal's or a human's right not to be killed may justifiably be violated. If Cochrane wants to simultaneously hold that certain beings, on the one hand, have a much weaker right not to be killed than others, but, on the other hand, still possess an equally strong right not to be killed in many concrete situations, for instance, experimentation, then Cochrane allegedly interprets situations of self-defense more or less broadly, depending first and foremost on the prudential unity relations of the individual whose right not to be killed is at stake.

I desperately hope that in no way I have given the impression of supporting lethal experimentation on mentally disabled human beings or the abolishment of the universal human right not to be killed. I acknowledge that there are indeed very good reasons against regarding the human right not to be killed as a matter of degree and why this fundamental human right should not be tampered with. Similarly, there are very good reasons against using cognitively disabled human beings or small infants for lethal experiments, just like there may be very good reasons not to use animals. Yet importantly, these reasons may not necessarily be the same. To attempt to justify an animal's right not to be killed in an experiment by reference to the equivalent right of a cognitively disabled human being by arguing that because, as a matter of fact, we do not use cognitively disabled humans in experiments, this showed that obviously their only weak time-relative interest in continuing to live was sufficient for ascribing them with this strong right, is neither promising, nor convincing, in my view.

CHAPTER 7

Final Conclusion

> So many veils and illusions separate us from the stark knowledge that we are dying; when we finally know we are dying, and all other sentient beings are dying with us, we start to have a burning, almost heartbreaking sense of the fragility and preciousness of each moment and each being, and from this can grow a deep, clear, limitless compassion for all beings. (Rinpoche 2012, 191)[154]

This thesis has been concerned with the question whether animals are harmed through their deaths, that is, whether they have an interest in continuing to live. The easy, short answer to this question is affirmative: yes, all sentient animals have an interest in continuing to live and are harmed through their deaths. This interest, however, plausibly occurs in degrees in that some individuals have a stronger interest in continuing to live than others. According to Jeff McMahan's time-relative interest account, which has been argued for in chapters 2 and 3, the strength of this interest, first and foremost, varies with the degree of the so-called prudential unity relations, that is, those psychological relations that bind an individual to her future self.

Based on the analyses and considerations that have been made throughout this thesis, all those very different kinds of sentient animals (and humans[155]) that can be harmed through their deaths can be roughly classified into four groups. The first of these four groups, which will be referred to as group 0, contains psychologically unconnected beings, that are harmed through their deaths but only in an absolutely minimal sense. In chapter 3.4.1, I suggested that it is at least questionable whether there are any real-life 'Psucos' in the animal kingdom. At least those animals in the focus of an ethics of killing – vertebrates and also some invertebrates, for instance, cephalopods – most certainly do not fall into the category of sentient, psychologically completely unconnected beings (see also Simmons 2010, 390, fn 37).

154 Sogyal Rinpoche attributes the above quote to Sir Thomas More.

155 As this thesis has been concerned with *animals'* interest in continuing to live, I will mainly focus on animals hereafter as well. Yet, the four groups apply to all kinds of sentient beings and thus naturally include human beings as well.

 | DOI:10.30965/9783957437365_008

Group 1 contains all those animals who can form direct psychological connections into their future. These connections, however, only concern mental states that are being retained through time (retention-typed connections), and internal reference-typed connections into the very immediate future (see chapter 3.3.2 for details). Animals from this group can make certain 'predictions' about events that are yet to come, but they do not possess an actual sense of the future in terms of its potential alterability. Their knowledge of what is going to happen is probably based on a script-based knowledge that enables them to constantly update their model of the world, just like scripts allow young children to know that because they have just brushed their teeth, it is time for bedtime-story-reading next. The subjects of this category may also engage in mental space travel, causing them to act in a future-oriented kind of way, but in the absence of any grasp of the future, mental space travel here functions more like a 'trigger' of the goal-oriented behavior in the sense described in chapter 4.3.2 and does not yet indicate any planning abilities.

The capacity to engage in anticipatory planning represents the transition from animals of group 1 to group 2. At this stage, a grasp of temporal, rather than aspectual notions, awakens. Future events are now being represented as alterable in contrast to past, unalterable events. As animals capable of anticipatory planning know that their future needs and desires can be different from their present ones, it seems sensible to attribute them with a grasp of their own future. Animals of this group might plausibly be classified as also fulfilling Locke's criterion of a person, that is, as beings aware of themselves as distinct entities with a past and a future. (Note that this interpretation of Lockean self-consciousness is much more demanding than Singer's, which has been presented in chapters 6.1 and 6.4.1.) Because animals of group 2 can be rationally ascribed with a sense of the future, their mental space travel ability now in fact represents a form of future thinking. In chapter 6.3.5.3, it was furthermore argued that the capability of anticipatory planning should be regarded as fulfilling the criteria of McMahan's rudimentary autonomy, so that animals of group 2 should be regarded as rudimentarily autonomous.

Group 3, finally, includes all those beings, whose temporal perspective goes one step further in making the transition from event-dependent to event-independent thinking abilities. This event-independent way of thinking constitutes itself, for example, in temporal-causal reasoning. The temporal perspective at this stage is required for strategic planning. Subjects with strategic planning capabilities can simulate the future by ways of mental time travel, now, and have the capacity to feel (anticipated) regret. As has become evident in chapter 5.2, there seems to be no conclusive evidence for strategic planning abilities in animals at present. In sum, therefore, it may be the case that most (or even all) sentient animals either fall into group 1 or group 2.

To kill an animal with anticipatory planning abilities, that is, an animal from group 2, means to brutally cut all those connections reaching into her future. In fact, it was argued in chapter 3.3.2 that such an animal should indeed be ascribed with a strong degree of psychological unity in McMahan's sense and thus, *ceteris paribus*, with a strong interest in continuing to live. On an interest-based rights account like Cochrane's it is plausible to ascribe these animals with a strong right not to be killed that should in fact only be overridden in very severe cases, analogous to situations of self-defense.

Yet, in light of current research and the conceptual clarifications that have been made throughout this thesis, I have argued that only very few animal species thus far seem to be capable of planning at all. The vast majority of the animal kingdom very probably falls into the category of group 1[156], in contrast, encompassing beings lacking any sense of the future in terms of its alterability. On the assumption that McMahan's time-relative-interest account is correct, death may very well not represent a major harm to these animals, (at least in comparison to the harm that death does to those beings subdued under group 3). There may, of course, be other very good reasons not to kill these animals, but if the considerations in this thesis have been correct, these additional reasons do not determine the harm done to the animals themselves.

Let me come back to the example of the happy pig Edward, that was made in the beginning of this thesis. The claim that in painlessly ending pig Edward's life, presumed he is a member of group 1, the farmer only does a minor (or at least not a very strong) harm to the animal, will be certainly unbearable to many, especially for those with a concern for animal welfare (and here I include myself). And indeed, from an anthropocentric point of view, the idea that death might not always represent one of the worst harms that can be done to an individual is very hard to grasp. So how should we proceed?

One logical possibility would be to attempt to save the pro-animal attitude that death strongly harms all sentient animals in arguing, for instance, that (McMahan's) time-relative interest account of the badness of death is simply incorrect.[157]

---

156 Of course, it would be sensible to further divide this large group of animals into several other subgroups. The focus of my thesis has been on the transitions of group 1 to group 2 and group 3, however, so this task must be reserved for other work.

157 In fact, many authors, among them Višak, Cochrane and DeGrazia, criticize McMahan's time-relative interest account for determining an individual's loss of future good as an objective parameter, when it is much more sensible to evaluate this loss from the subject's perspective herself, relative to what the individual could have maximally reached due to

If, in contrast, the time-relative interest account and the above considerations are being accepted, then the conclusion that most animals are comparably only weakly harmed through their deaths must be made meaningful. Below, one possible solution how this could be achieved will be proposed. In doing so, I will suggest that the reason why we as paradigmatic human adults are so reluctant to accept that death may not do a terrible harm to some individuals, might lie in the fact that we tend to project our own fear of death onto other beings, among them animals. On the first glance, this admittedly speculative claim may sound unconvincing or even blatantly absurd, so I will attempt to present a sensible argument in the following paragraphs.

Human adult persons probably are the only beings on earth, who do not only abstractly conceive of their mortality in that they understand that all living things will die someday (which other, nonhuman animals may be capable of as well), but who can literally imagine a time after their own deaths. Having a grasp of time as infinitely flowing from past to present and into the future, persons do not only represent time in an event-independent sense but also an *I-independent* sense, so to speak; they understand that their deaths do not represent the ending of the world and of time, but that life will simply go on without them.

That our lives are highly affected by our knowledge of mortality is not a novel thought and has been expressed in numerous philosophical and other writings. For a long time, however, this idea remained a theoretical hypothesis and was even considered to be empirically unverifiable. Throughout the last 25 years, however, it has been put on stable empirical ground (for an overview, see Solomon et al. 2015). Inspired by existentialist philosophy and, first and foremost, Ernest Becker's Pulitzer Prize-winning book *The Denial of Death*, the three psychologists Jeff Greenberg, Sheldon Solomon and Tom Pyszczynski developed the *terror management theory* in the early 1990s.

In a nutshell, this theory proposes that many of our behaviors in virtually all areas of life are attributable to two self-defense strategies, coming into play whenever the truth of death's inevitable and unpredictable nature is threatening to overwhelm us. If the idea of death is conscious to us, for instance, during the funeral of a family member or after witnessing a car accident, the *proximal defense* is active: we are trying to distract ourselves from the inconvenient truth of our own mortality and attempt to get the uncomfortable thought as quickly as possible out of our heads ("*not me, not now*"). *Distal defenses*

---

her specific cognitive architecture. In chapter 3.2.2, I have laid out certain problems such a modification of the TRIA faces. This is not to argue against such a modification per se, of course, nor do I want to argue that these challenges cannot be overcome.

respond to an unconscious, more subtle presence of death in our minds. First and foremost, they cause a boost in self-esteem and cultural identity, preserving precious values and beliefs. The purpose behind these distal defense strategies thus seems to be to prevent us from despairing at the alleged insignificance of our lives. We want to achieve something in life to give our existence a symbolic meaning, reaching beyond our physical deaths (Solomon et al. 2015, 171ff).

Children know about their mortality even at a very young age, but this knowledge remains on a rather abstract level for quite a long time. It is only during adolescence that this abstract knowledge grows into the angst-inducing certainty that death, and, most importantly, one's *own* death, is inevitable and its timing unpredictable, which makes death omnipresent in the back of our minds. Solomon et al. (2015) do not speak for nothing of a *terror* of death rather than just a fear.

In light of this terror, the countless occasions, in which philosophers and others have denied animals any bit of future thinking capabilities, may gain a slightly different meaning. Consider, once again, the following observation by Friedrich Nietzsche that has already been quoted earlier:

> Consider the cattle, grazing as they pass you by: they do not know what is meant by yesterday or today, they leap about, eat, rest, digest, leap about again, and so from morn till night and from day to day, fettered to the moment and its pleasure or displeasure, and thus neither melancholy nor bored. This is a hard sight for man to see; for, though he thinks himself better than the animals because he is human, he cannot help envying them their happiness. (Nietzsche 1876/1983, 60f; cited in Hoerl 2008,486)

The envy that Nietzsche alludes to becomes especially plausible in light of terror management theory. For human persons, great happiness may, explicitly or implicitly, be accompanied by a certain melancholy because of the fleetingness of time. A linear conception of time implies the knowledge that every unique moment in time will be irreversibly lost once it has passed. The quotation at the beginning of this chapter speaks of a "burning, almost heartbreaking sense of the fragility and preciousness of each moment and each being" (Rinpoche 2012, 191). We cannot only regret (*Bereuen*) false decisions but also feel regret or sorrow (*Bedauern*) in light of positive experiences lying in the past. These considerations resemble Søren Kierkegaard's remarks on the human fear of death, which he termed the *sickness unto death* (Kierkegaard 1849/2005). For Kierkegaard the human person is an indissoluble synthesis of immortality and mortality, which cannot be dissolved, but to which she necessarily must put herself into *some* sort of relation.

Under this reading, the widespread view that children and animals experience somehow 'purer' moments of happiness, might become more meaningful, because positive experiences should indeed be very different, if they are not dampened by an awareness of (one's own) mortality. In this specific sense, children and animals can *in fact* be viewed as living in the moment; not necessarily because they cannot anticipate any future desires or needs or sacrifice present pleasures in favor of future outcomes, but because their present joy is not diminished by the knowledge of mortality and irreversibility. All sentient beings can experience pleasures and some of them can undoubtedly feel also happiness and joy, but they may be doing so in ways very different to ours.

On the one hand, not only to be alive, but to be *aware* that one is, can cause a deep thankfulness and may, thus, contribute to the happiness felt in that particular moment in human adult persons. On the other hand, the consciousness of existence, together with the abstract knowledge of one's mortality also reveals a tragedy; yes, we know that we are alive, but we also know that we will not always be.

One way to cope with the uncomfortable awareness of mortality could be to simply deny it, which could be argued to be the prevailing strategy of today's Western societies. Just like aging and death have been pushed to the margins of society so that we do not have to deal with the thought of our own and our loved ones' deaths, the killing of animals in the process of meat-production (and also experimentation) has been very carefully hidden from the eyes of the consumers. Of course, most omnivores, with the possible exception of young children, know that meat is animal flesh. In accordance with Solomon et al. (2015)'s two defense strategies, however, they might attempt to deny this knowledge by, for example, choosing not to think about it or by consuming meat in forms that only slightly resemble flesh.

Alternatively, omnivores might try to broaden the alleged gap between humans (*us*) and all other animals (*them*), that is, emphasize the anthropological difference (Wild 2010). In emphasizing both the in-group identity and feeling of solidarity among all human beings as well as the postulated vast differences between humans and animals, this justifies killing and eating *them* but not *one of us*.

Experimental support for the above hypotheses comes from Uri Lifshin et al. (2017) as well as Goldenberg et al. (2001), among others, who found that people, who had been reminded of their mortality, have an increased need to distance themselves from animals and tend to support the killing of animals more than controls.[158] According to Marino and Mountain (2015) the fear of

158 Importantly, other negative thoughts, e.g. of failure, future stressful events, pain or mutilation did not have similar effects.

death lies at the basis of the exploitive and harmful nature of our relationship towards all other animals. Because this relationship has so many negative consequences, not only for the millions of animals, suffering and killed from our hands, but also for the survival of humankind (challenged by mass extinctions, ocean pollution, climate change etc.), the authors argue that "the only viable future for humankind lies in achieving some level of acceptance of our own mortality and developing a more humble, and ultimately more satisfying, relationship to our fellow animals and the natural world" (Marino and Mountain 2015, 18). If, in a nutshell, we want to cure our relationship towards other animals, we must first cure the relationship towards our own mortality.

Many vegetarians (and also omnivores!) now seem to think that a thorough and honest analysis of our fear of death and our relationship towards our own mortality automatically leads to vegetarianism or even veganism. Once a person allows herself to actually face the truth that the meat on her plate once was a breathing, living and mortal creature just like she is, once she develops a "deep, clear, limitless compassion for all beings" (Rinpoche 2012, 191), she cannot but conclude that eating animals is extremely wrong. If this was true, then only care- and thoughtless consumers would be continuing to eat meat, either by carefully denying that they are eating the flesh of a once-alive animal or by emphasizing alleged differences between humans and all other animals. Omnivorism would always be the expression of an underlying denial on the part of these consumers, on this view; either a denial of their own mortality or a denial of their animality.

Even though I certainly think that many, probably even the large majority of omnivores, indeed either successfully deny the causal connection between the tasty meat and the breathing animal or alternatively aim to justify their gustatory preferences by downgrading the moral status of farm animals while simultaneously upgrading their own species' capabilities – similarly to what has been described as anthropofabulation (Buckner 2013) in a different context –, I do not believe that an acceptance of one's own mortality and the cure of our exploitive relationship towards animals must *necessarily* lead to a plant-based diet. Instead, because death has been pushed to the outer margins of our society, we seem to have lost a 'healthy', not overly-emotional view on death, which may sometimes prevent us from rationally determining its harm for other beings.

Let me clarify this claim. The home slaughtering of animals in the presence of children was common in rural Germany until the 1980s and still is in preindustrial cultures. Even though children normally did not participate in the killing itself, they were allowed to watch and also took part in the subsequent cutting and processing of the carcass. Ever since, death and physical violence have been pushed to the outer margins of society, however, and the production of meat has become mechanized in the slaughter houses, so that home slaughtering in industrial nations has become very rare. As a result, the killing of animals has become a social taboo, inducing fear and unease on behalf of adults (Rose 2015, 231f).

In dealing with the historical, cultural and psychological reasons for why the slaughtering of a rabbit in the presence of school children could cause a nationwide medial outrage several years ago in Germany, the educationist Lotte Rose argues that in their wish to protect their children, parents actually projected their own fears onto their children. The children were said to need protection from a potentially traumatizing event, when in fact it was the *adults* who lacked confidence that they would be able to cope with the violent event.[159] Adults, Rose argues, project their own fear of the dead animal onto the "vulnerable child", even though this may prevent the child from developing a rational, not overly-emotional view of death.

In light of terror management theory, Rose's argument could be taken one step further in postulating that it is, first and foremost, the adult fear of death and the terror of one's own mortality that is being projected onto children. In preventing children from witnessing the killing of an animal, parents also express the wish to protect their children from the awareness of their own mortality, which they know will awaken sooner or later.

If it is correct that we tend to project our own fear of death onto other beings, this raises the question whether some people might project this fear onto other feeling and thinking beings as well, for instance, animals. Vegetarianism and veganism is particularly widespread among a young, well-educated (mostly female) population living in urban areas, that is, among human beings, who have only rarely had personal encounters with farm animals, more so, if they grew up in urban environments as well. It might be argued that especially these self-proclaimed animal-loving, but occasionally rather inexperienced individuals, having grown up in a culture, that has successfully eliminated death from its social center, might be at risk of irrationally projecting death's terror onto animals.[160]

True, someone deciding to become a vegetarian or vegan on moral grounds no longer denies the causal connection between the meat on the plate and the

159 Of course, the slaughtering of an animal *can* indeed be traumatizing for children, even more so, if the animal shows signs of pain, fear and distress. Rose cautions, however, to generalize the experience of few individuals onto the practice as such (Rose 2015, 239).

160 To be clear, I do not want to make the claim that vegetarians or vegans *per se* are projecting the terror of death onto animals, nor do I want to argue that vegetarianism or veganism are always expressions of anthropomorphism, just as I do not want to argue that meat-eaters *per se* are insensitive and cruel.

dead animal. She does not automatically gain a more *truthful* view on the significance of death for animals and the killing of animals, however. Instead of trying to *broaden* the gap between humans and animals – as many meat eaters have been empirically shown to do – these persons could be said to strive to *close* it. Both options might have little to do with reality and should be rejected, in my opinion.

To claim, on the one hand, that animals are not at all harmed through their deaths is plainly wrong, both in light of the time-relative interest account as well as the empirical findings that have been discussed. On the other hand, it is equally wrong to defend the claim that all sentient beings have an equally strong interest in continuing to live, in my view. Indeed, as has been argued throughout this thesis, a time-relative interest account along the lines of McMahan should come to the conclusion that death only does a minor harm to those animals incapable of forming internal reference-typed connections into the non-immediate future.

How this minor harm is to be weighed regarding its ethical evaluation is a different story. An interest-based rights account like Cochrane's has certain major advantages over utilitarian and deontological approaches, because it is able to protect its bearers from being all too easily sacrificed in favor of rivaling interests and because it acknowledges that the right not to be killed is a matter of degree. If the postulation of a gradual right not to be killed is taken seriously, however, then this must also be mirrored in the degree of moral protection in concrete examples of use.[161]

Consequently, animals from group 2, in possession of anticipatory planning capabilities, should be ascribed with a much stronger moral right not to be killed than animals from group 1, unable to form any internal reference-typed connections into the non-immediate future. That means that there are plausibly more and/or other situations, in which a right not to be killed can justifiably be violated in the case of those animals belonging to group 1 in contrast to those from group 2. How exactly these situations should be defined, whether they only apply, for instance, to cases of self-defense or are to be defined more broadly, cannot be easily answered and requires a much more thorough analysis than can be provided here.

---

161 In contrast, an approach simultaneously holding that the right not to be killed can be more or less strong *and* that a weak and a strong right have similar practical implications, is at least obliged to plausibly justify this move. In doing so, it is not sufficient, in my opinion, to argue that our refusal to raise mentally disabled human beings for meat or lethal experiments allows to make direct inferences to the strength of their moral status.

I believe, however, that it is indeed plausible to argue that animals with planning capabilities should only be allowed to be killed under very exceptional cases of self-defense, similar to Regan's lifeboat scenarios. With regard to those animals from group 1, who can, for example, retain desires or beliefs through time and who might also have the ability to engage in mental space travel, but who are unable to form any internal reference-typed connections into the non-immediate future, the actual weight that their interest in continuing to live should have, is much more dubious, I believe, especially if a certain animal only has a very small number of direct psychological connections into the future in total.

Of course, there are countless other reasons in favor of vegetarianism or veganism, which are not dependent on the harm that death does to an animal, such as environmental, humanitarian or health-based ones, to name just a few. Countless books, papers and articles have been written on the manifold benefits of a plant-based diet, so I will merely allude to one, rather practical reason here: most informed (i.e., non-denying) omnivores would probably agree that the practice of factory farming should be abandoned as quickly as possible and many might also acknowledge that consuming products that come with so much suffering and pain is morally wrong. Many of us should agree, therefore, that consumers should certainly avoid meat and other animal products from factory farms where possible. This, however, is much more difficult for omnivores than it is for vegans or vegetarians, at least in urban areas. It is not always possible to know how a particular piece of meat has been produced, for example in a restaurant, nor is it always an option to ask for this information, for instance, when attending your boss's dinner or the kindergarten's charity bazaar.

Furthermore, it is entirely unclear whether the idea this whole philosophical inquiry rests upon, that is, the possibility of killing an animal without inflicting any kind of pain, suffering, fear or other forms of distress, in fact is a practicable option or remains nothing more than a theoretical thought-experiment. And finally, as has repeatedly been stated, future research might very well indicate planning abilities in pigs and other animals commonly used for animal agriculture as well. Until then, it might thus indeed be better to err on the side of caution. All in all, the risk of involuntarily violating an animal's interest (or right) not to be made to suffer or even not to be killed is much higher for an omnivore than it is for a vegetarian or vegan.

In sum, to accept that many animals might not be strongly harmed through their deaths, is not to deny that there may not be very good reasons for not killing them. I certainly acknowledge that this may still be an unwelcome conclusion for many animal activists and ethicists. At the same time, however, future research will continue to explore and verify future thinking capabilities in a lot more species than in the few currently known of. To what extent animals may be capable of prospective cognition and planning remains to be one of the most active and hotly contested questions in the philosophy of animal minds and comparative cognition, so the future will undoubtedly provide us with more astonishing findings.

Bibliography

Ach, J. S. (2009). Transgene Tiere: Anmerkungen zur Herstellung, Nutzung und Haltung transgener Tiere aus tierethischer Perspektive. In J. S. Ach & M. Stephany (Eds.): *Die Frage nach dem Tier: Interdisziplinäre Perspektiven auf das Mensch-Tier-Verhältnis* (pp. 33–45). Berlin: LIT.

Atance, C. M., & O'Neill, D. K. (2001). Episodic future thinking. *Trends in Cognitive Sciences 5*(12), 533–539.

Atance, C. M., & O'Neill, D. K. (2005). The emergence of episodic future thinking in humans. *Learning and Motivation 36*(2), 126–144.

Augustinack, J. C., van der Kouwe, A. J. W., Salat, D. H., Benner, T., Stevens, A. A., Annese, J., … Corkin, S. (2014). H.M.'s contributions to neuroscience: a review and autopsy studies. *Hippocampus 24*(11), 1267–1286.

Bäuerlein, T. (2011). *Fleisch essen, Tiere lieben: Wo Vegetarier sich irren und was Fleischesser besser machen können*. Munich: Ludwig.

Beauchamp, T. L. (2005). Who Deserves Autonomy, and Whose Autonomy Deserves Respect. In J. S. Taylor (Ed.), *Personal Autonomy: New Essays on Personal Autonomy and its Role in Contemporary Moral Philosophy* (pp. 310–329). Cambridge University Press.

Beauchamp, T. L., & Wobber, V. (2014). Autonomy in Chimpanzees. *Theoretical Medicine and Bioethics 35*(2), 117–132.

Beck, S. R., Robinson, E. J., Carroll, D. J., & Apperly, I. A. (2006). Children's Thinking About Counterfactuals and Future Hypotheticals as Possibilities. *Child Development 77*(2), 413–426.

Becker, E. (1973). *The Denial of Death*. New York: Free Press.

Belshaw, C. (2015). Death, Pain, and Animal Life. In P. Singer, T. Višak & R. Garner (Eds.), *The Ethics of Killing Animals* (pp. 32–50). Oxford University Press.

Bentham, J. (1970). *An introduction to the principles of morals and legislation*. London: Athlone Press.

Beran, M. J., Parrish, A. E., Futch, S. E., Evans, T. A., & Perdue, B. M. (2015). Looking ahead? Computerized maze task performance by chimpanzees (Pan troglodytes), rhesus monkeys (Macaca mulatta), capuchin monkeys (Cebus apella), and human children (Homo sapiens). *Journal of Comparative Psychology 129*(2), 160–173. Author manuscript.

Bird, C. D., & Emery, N. J. (2009). Insightful problem solving and creative tool modification by captive nontool-using rooks. *Proceedings of the National Academy of Sciences of the United States of America 106*(25), 10370–10375.

Birnbacher, D. (2009). Haben Tiere Rechte? In J. S. Ach & M. Stephany (Eds.), *Die Frage nach dem Tier: Interdisziplinäre Perspektiven auf das Mensch-Tier-Verhältnis* (pp. 47–64). Berlin: LIT.

Boeckle, M., & Clayton, N. S. (2017). A raven's memories are for the future. *Science 357*(6347), 126–127.

Boyer, P. (2008). Evolutionary economics of mental time travel? *Trends in Cognitive Sciences 12*(6), 219–224.

Bradley, B. (2015). Is Death Bad for a Cow? In P. Singer, T. Višak & R. Garner (Eds.), *The Ethics of Killing Animals* (pp. 51–64). Oxford University Press.

Bradley, B., Feldman, F., & Johansson, J. (2013). *The Oxford Handbook of Philosophy of Death*. Oxford University Press.

Bräuer, J., & Call, J. (2015). Apes produce tools for future use. *American Journal of Primatology, 77*(3), 254–263.

Broome, J. (2013). The Badness of Death and the Goodness of Life. In B. Bradley, Feldman, F., & Johansson, J. (Eds.), The Oxford Handbook of Philosophy of Death (pp. 218–233). Oxford University Press.

Bruce, M., & Barbone, S. (2011). *Just the Arguments: 100 of the Most Important Arguments in Western Philosophy*. Chichester: Wiley-Blackwell.

Buckner, C. (2013). Morgan's Canon, meet Hume's Dictum: avoiding anthropofabulation in cross-species comparisons. Biology & Philosophy, 28(5), 853–871.

Buckner, C. (2018). Understanding associative and cognitive explanations in comparative psychology. In *The Routledge Handbook of Philosophy of Animal Minds* (pp. 409–418). London, New York: Routledge.

Buckner, R. L., & Carroll, D. C. (2007). Self-projection and the brain. *Trends in Cognitive Sciences, 11*(2), 49–57.

Budde, E. T. (2015). *Abtreibungspolitik in Deutschland: Ein Überblick*. Wiesbaden: Springer Fachmedien.

Buss, S. (2014). Personal Autonomy. In E. N. Zalta (Eds.), *The Stanford Encyclopedia of Philosophy* (Winter 2014). Retrieved from https://plato.stanford.edu/archives/win2014/entries/personal-autonomy/.

Campbell, J. (2006). Ordinary Thinking About Time. In M. Stöltzner & F. Stadler (Eds.), *Time and History: Proceedings of the 28. International Ludwig Wittgenstein Symposium, Kirchberg Am Wechsel, Austria 2005* (pp 1–12). Berlin: De Gruyter.

Cavalieri, P. (2002). *Die Frage nach den Tieren. Für eine erweiterte Theorie der Menschenrechte*. Erlangen: Fischer.

Cheke, L. G., & Clayton, N. S. (2010). Mental time travel in animals. *Wiley Interdisciplinary Reviews: Cognitive Science 1*(6), 915–930.

Cheke, L. G., & Clayton, N. S. (2012). Eurasian jays (Garrulus glandarius) overcome their current desires to anticipate two distinct future needs and plan for them appropriately. *Biology Letters 8*(2), 171–175.

Cheng, S., & Werning, M. (2016). What is episodic memory if it is a natural kind? *Synthese, 193*(5), 1345–1385.

Clayton, N. S., Correia, S. P. C., Raby, C. R., Alexis, D. M., Emery, N. J., & Dickinson, A. (2008). Response to Suddendorf & Corballis (2008): in defence of animal foresight. *Animal Behaviour 4*(76), e9–e11.

Cochet, H., & Byrne, R. W. (2015). Complexity in animal behaviour: towards common ground. *Acta Ethologica 18*(2), 237–241.

Cochrane, A. (2007). Animal Rights and Animal Experiments: An Interest-Based Approach. *Res Publica 13*(3), 293–318.

Cochrane, A. (2012). *Animal Rights Without Liberation: Applied Ethics and Human Obligations*. New York: Columbia University Press.

Cochrane, A. (2013). Cosmozoopolis: The Case Against Group-Differentiated Animal Rights. *Law, Ethics and Philosophy, 1*(0), 127–141.

Cochrane, A. (2015). Life, Liberty, and the Pursuit of Happiness? Specifying the Rights of Animals. In T. Višak & R. Garner (Eds.), *The Ethics of Killing Animals* (pp. 201–214). Oxford University Press.

Cohen, C. (2001). Reply to Tom Regan. In C. Cohen & T. Regan (Eds.), *The Animal Rights Debate* (pp. 225–261). Lanham: Rowman & Littlefield.

Cohen, C., & Regan, T. (2001). *The Animal Rights Debate*. Lanham: Rowman & Littlefield.

Corballis, M. C. (2013). The Wandering Rat: Response to Suddendorf. *Trends in Cognitive Sciences 17*(4), 152–152.

Cordonnier, A., Barnier, A. J., & Sutton, J. (2016). Scripts and information units in future planning: Interactions between a past and a future planning task. *The Quarterly Journal of Experimental Psychology* 69(2), 324–338.

Corkin, D. S. (2014). *Permanent Present Tense: The man with no memory, and what he taught the world.* London: Penguin.

Correia, S. P. C., Dickinson, A., & Clayton, N. S. (2007). Western scrub-jays anticipate future needs independently of their current motivational state. *Current Biology 17*(10), 856–861.

Craver, C. F., Kwan, D., Steindam, C., & Rosenbaum, R. S. (2014). Individuals with episodic amnesia are not stuck in time. *Neuropsychologia 57*, 191–195.

D'Argembeau, A., Renaud, O., & Linden, M. V. der. (2011). Frequency, characteristics and functions of future-oriented thoughts in daily life. *Applied Cognitive Psychology 1*(25), 96–103.

Darwin, C. (1871/2004). *The descent of man, and selection in relation to sex*. London: Penguin.

Davidson, D. (1982). Rational Animals. *Dialectica 36*, 317–28.

Davis, S. L. (2003). The Least Harm Principle May Require that Humans Consume a Diet Containing Large Herbivores, Not a Vegan Diet. *Journal of Agricultural and Environmental Ethics 16*(4), 387–394.

DeGrazia, D. (1996). *Taking Animals Seriously: Mental Life and Moral Status*. Cambridge University Press.

DeGrazia, D. (2007). The Harm of Death, Time-Relative Interests, and Abortion. *The Philosophical Forum 38*(1), 57–80.

Donaldson, S., & Kymlicka, W. (2013). *Zoopolis*. Oxford University Press.

Dworkin, G. (1988). *The Theory and Practice of Autonomy*. Cambridge University Press.

Eacott, M. J., & Easton, A. (2012). Remembering the past and thinking about the future: Is it really about time? *Learning and Motivation 43*(4), 200–208.

Eck, S. B. (2002). Auf dem Prüfstand: Albert Schweitzer und die Ethik der Ehrfurcht vor dem Leben. In Arbeitskreis Tierrechte und Ethik (Ed.), *Retrospektive*. Saarbrücken: Arbeitskreis Tierrechte und Ethik.

Edwards, R. B. (1993). Tom Regan's Seafaring Dog and (Un)Equal Inherent Worth. *Between the Species 9*(4), 9.

Fabre, J. H. (1915/2001). *The hunting wasps*. New York: Dodd, Mead and Company.

Fitzpatrick, S. (2009). The Primate Mindreading Controversy: A Case Study in Simplicity and Methodology in Animal Psychology. In R. W. Lurz (Ed.), *The Philosophy of Animal Minds* (pp. 224–246). Cambridge University Press.

Fodor, J. (1999). Not so Clever Hans. *London Review of Books* (February 4, 1999), 12–13.

Foer, J. S. (2009). *Eating animals*. London: Hamilton.

Frankfurt, H. G. (1971). Freedom of the Will and the Concept of a Person. *Journal of Philosophy 68*(1), 5–20.

Fugazza, C., Pogány, Á., & Miklósi, Á. (2016). Recall of Others' Actions after Incidental Encoding Reveals Episodic-like Memory in Dogs. *Current Biology 26*(23), 3209–3213.

Garner, R. (2013). *A Theory of Justice for Animals: Animal Rights in a Nonideal World*. Oxford University Press.

Garner, R. (2015). Welfare, Rights, and Non-Ideal Theory. In T. Višak & R. Garner (Eds.), *The Ethics of Killing Animals* (pp. 215–228). Oxford University Press.

German, T. P., & Nichols, S. (2003). Children's counterfactual inferences about long and short causal chains. *Developmental Science 6*(5), 514–523.

Gilbert, D. T., & Wilson, T. D. (2007). Prospection: Experiencing the Future. *Science 317*(5843), 1351–1354.

Goldenberg, J. L., Pyszczynski, T., Greenberg, J., Solomon, S., Kluck, B., & Cornwell, R. (2001). I am not an animal: mortality salience, disgust, and the denial of human creatureliness. *Journal of Experimental Psychology. General 130*(3), 427–435.

Gottlieb, J. (2014). Vegetarianism. Interview with Peter Singer and Jeff McMahan. *Minerva Podcast, Episode 15* (May 31, 2014). Retrieved from http://www.minerva-podcast.com/post/87392358458/veg. (Jan 3, 2017).

Grimm, H., & Wild, M. (2016). *Tierethik zur Einführung*. Hamburg: Junius.

Harman, E. (2011). The Moral Significance of Animal Pain and Animal Death. In T. L. Beauchamp & R. G. Frey (Eds.), *The Oxford Handbook of Animal Ethics* (pp. 726–737). Oxford University Press. Author manuscript.

Hassabis, D., & Maguire, E. A. (2009). The construction system of the brain. *Philosophical Transactions of the Royal Society of London B: Biological Sciences 364*(1521), 1263–1271.

Hayes-Roth, B., & Hayes-Roth, F. (1979). A Cognitive Model of Planning. *Cognitive Science 3*(4), 275–310.

Hesslow, G. (2002). Conscious thought as simulation of behaviour and perception. *Trends in Cognitive Sciences 6*(6), 242–247.

Heyes, C. M. (1998). Theory of mind in nonhuman primates. *Behavioral and Brain Sciences 21*(1), 101–114.

Heyes, C. (2012). Simple minds: a qualified defence of associative learning. *Philosophical Transactions of the Royal Society of London. Series B, Biological Sciences 367*(1603), 2695–2703.

Hoerl, C. (2008). On Being Stuck in Time. *Phenomenology and the Cognitive Sciences 7*(4), 485–500.

Hoerl, C., & McCormack, T. (2011). Time in Cognitive Development. In C. Callender (ed.), *The Oxford Handbook of Philosophy of Time* (pp. 439–459). Oxford University Press.

Hoerl, C., & McCormack, T. (2016). Making decisions about the future: regret and the cognitive function of episodic memory. In K. Michaelian, S. B. Klein, & K. Szpunar (Eds.), *Seeing the Future: Theoretical Perspectives on Future-Oriented Mental Time Travel* (pp. 241–266). Oxford University Press.

Horta, O. (2014). The Scope of the Argument From Species Overlap. *Journal of Applied Philosophy 31*(2), 142–154.

Inside the Animal Mind: The Problem Solvers. G. Russell. C. Packham. BBC Two, 2014. Retrieved from https://www.youtube.com/watch?v=cbSu2PXOTOc (January 12, 2018).

Jamieson, D. (1990). Rights, Justice, and Duties to Provide Assistance: A Critique of Regan's Theory of Rights. *Ethics 100*(2), 349–362.

Jelbert, S. A., & Clayton, N. S. (2017). Comparing the non-linguistic hallmarks of episodic memory systems in corvids and children. *Current Opinion in Behavioral Sciences 17*, 99–106.

Kabadayi, C., & Osvath, M. (2017). Ravens parallel great apes in flexible planning for tool-use and bartering. *Science 357*(6347), 202–204.

Kagan, S. (2015). Singer on Killing Animals. In T. Višak & R. Garner (Eds.), *The Ethics of Killing Animals* (pp. 136–153). Oxford University Press.

Keijzer, F. (2013). The Sphex story: How the cognitive sciences kept repeating an old and questionable anecdote. *Philosophical Psychology 26*(4), 502–519.

Kierkegaard, S. (1849/2005). *Die Krankheit zum Tode*. Munich: Deutscher Taschenbuch Verlag.

Klein, S. B. (2016). Autonoetic consciousness: Reconsidering the role of episodic memory in future-oriented self-projection. *Quarterly Journal of Experimental Psychology 69*(2), 381–401.

Klein, S., & Steindam, C. (2016). The role of subjective temporality in future-oriented mental time travel. In K. Michaelian, S. Klein, & K. Szpunar (Eds.), *Seeing the Future: Theoretical Perspectives on Future-Oriented Mental Time Travel* (pp. 135–152). Oxford University Press.

Korsgaard, C. (2006). Morality and the Distinctiveness of Human Action. In S. Macedo & J. Ober (Eds.), *Primates and Philosophers* (pp. 98–119). Princeton University Press.

Kwan, D., Craver, C. F., Green, L., Myerson, J., & Shayna, R. (2013). Dissociations in future thinking following hippocampal damage: Evidence from discounting and time perspective in episodic amnesia. *Journal of Experimental Psychology: General 142*(4), 1355–1369.

Lamey, A. (2007). Food Fight! Davis Versus Regan on the Ethics of Eating Beef. *Journal of Social Philosophy 38*(2), 331–348.

Li, H. (2002). Animal Research, Non-Vegetarianism, and the Moral Status of Animals: Understanding the Impasse of the Animal Rights Problem. *Journal of Medicine and Philosophy 27*(5), 589–615.

Lifshin, U., Greenberg, J., Zestcott, C. A., & Sullivan, D. (2017). The Evil Animal: A Terror Management Theory Perspective on the Human Tendency to Kill Animals. *Personality and Social Psychology Bulletin 43*(6), 743–757.

Locke, J. (1694/2008). Of Identity and Diversity. In J. Perry (Ed.), *Personal Identity* (pp. 33–52). Berkely, Los Angeles, London: University of California Press.

Luper, S. (2009). Death. In E. N. Zalta (Ed.), *The Stanford Encyclopedia of Philosophy* (Summer 2009). Retrieved from http://plato.stanford.edu/archives/sum2009/entries/death/.

Malanowski, S. (2016). Is episodic memory uniquely human? Evaluating the episodic-like memory research program. *Synthese 193*(5), 1433–1455.

Marino, L., & Mountain, M. (2015). Denial of Death and the Relationship between Humans and Other Animals. *Anthrozoös, 28*(1), 5–21.

Martin-Ordas, G., Schumacher, L., & Call, J. (2012). Sequential Tool Use in Great Apes. *PLOS ONE 7*(12), e52074, 1–15.

McColgan, K. L., & McCormack, T. (2008). Searching and planning: young children's reasoning about past and future event sequences. *Child Development 79*(5), 1477–1497.

McCormack, T. (2014). Three types of temporal perspective: characterizing developmental changes in temporal thought. *Annals of the New York Academy of Sciences 1326*(1), 82–89.

McCormack, T., & Hoerl, C. (2005). Children's Reasoning About the Causal Significance of the Temporal Order of Events. *Developmental Psychology 41*(1), 54–63.

McCormack, T., & Hoerl, C. (2007). Young children's reasoning about the order of past events. *Journal of Experimental Child Psychology 98*(3), 168–183.

McCormack, T., & Hoerl, C. (2011). Tool Use, Planning and Future Thinking in Children and Animals. In T. McCormack, C. Hoerl, & S. Butterfill (Eds.), *Tool use and causal cognition* (pp. 129–147). Oxford University Press.

McDermott, K. B., Wooldridge, C. L., Rice, H. J., Berg, J. J., & Szpunar, K. K. (2016). Visual perspective in remembering and episodic future thought. *The Quarterly Journal of Experimental Psychology 69*(2), 243–253.

McMahan, J. (1988). Death and the Value of Life. *Ethics 99*(1), 32–61.

McMahan, J. (1992). *Preferences, death, and the ethics of killing*. Conference manuscript. Retrieved from http://www.academia.edu/525192/Preferences_death_and_the_ethics_of_killing (March 20, 2014).

McMahan, J. (2002). *The Ethics of Killing: Problems at the Margins of Life*. Oxford University Press.

McMahan, J. (2005). Animals. In R. G. Frey & C. H. Wellman (Eds.), *A Companion to Applied Ethics* (pp. 525–536). Oxford: Blackwell.

McMahan, J. (2008). Eating animals the nice way. *Daedalus 137*(1), 66–76.

McMahan, J. (2010). The Meat Eaters. *The New York Times* (September 19, 2010). Retrieved from https://opinionator.blogs.nytimes.com/2010/09/19/the-meat-eaters/ (February 19, 2018).

McMahan, J. (2015). The Comparative Badness for Animals of Suffering and Death. In P. Singer, T. Višak & R. Garner (Eds.), *The Ethics of Killing Animals* (pp. 65–85). Oxford University Press.

Michaelian, K. (2017). Memory as mental time travel. *Routledge Handbook of Philosophy of Memory*.

Michaelian, K., Klein, S. B., & Szpunar, K. K. (2016). *Seeing the Future: Theoretical Perspectives on Future-Oriented Mental Time Travel*. Oxford University Press.

Michaelian, K & Sutton, J. (2017). Memory. In E. N. Zalta (Ed.), *Stanford Encyclopedia of Philosophy* (Summer 2017). Retrieved from https://plato.stanford.edu/archives/sum2017/entries/memory/.

Midgley, M. (1983). *Animals and why they matter*. Athens: University of Georgia Press

Mill, J. S. (1863/2008). Utilitarianism. Munich, Ravensburg: Grinverlag.

Morgan, C. L. (1903). *An introduction to comparative psychology*. London: Walter Scott Pub. Co.

Nietzsche, F. (1876/1983). *Untimely meditations*. Cambridge University Press.

Nozick, R. (1974/2008). *Anarchy, State, and Utopia*. New York: Basic Books.

Nozick, R. (1983). About Mammals and People. *New York Times Book Review (27 Nov 1983)*, 11.

Oliver, M. (1992). *New and Selected Poems* (*Vol. 1*). Boston: Beacon Press.

Osvath, M. (2009). Spontaneous planning for future stone throwing by a male chimpanzee. *Current Biology 19*(5), R190–R191.

Osvath, M. (2010). Great ape foresight is looking great. *Animal Cognition 13*(5), 777–781.

Osvath, M. (2016). Putting flexible animal prospection into context: escaping the theoretical box. *Wiley Interdisciplinary Reviews: Cognitive Science 7*(1), 5–18.

Osvath, M., & Karvonen, E. (2012). Spontaneous Innovation for Future Deception in a Male Chimpanzee. *PLOS ONE 7*(5), e36782, 1–8.

Osvath, M., & Martin-Ordas, G. (2014). The future of future-oriented cognition in non-humans: theory and the empirical case of the great apes. *Philosophical Transactions of the Royal Society B: Biological Sciences 369*(1655), 20130486, 1–8.

Osvath, M., & Osvath, H. (2008). Chimpanzee (Pan troglodytes) and orangutan (Pongo abelii) forethought: self-control and pre-experience in the face of future tool use. *Animal Cognition 11*(4), 661–674.

Parfit, D. (1987). *Reasons and persons*. Oxford: Clarendon Press.

Peckham, G. W. & Peckham, E. G. (1905). *Wasps, social and solitary*. Westminster: Constable.

Perry, J. (2008). *Personal identity*. Berkeley: University of California Press.

Pluhar, E. B. (1988). When is it morally acceptable to kill animals? *Journal of Agricultural Ethics 1*(3), 211–224.

Pluhar, E. B. (1995). *Beyond Prejudice: The Moral Significance of Human and Nonhuman Animals*. Durham: Duke University Press.

Pollan, M. (2007). *The Omnivore's Dilemma: A Natural History of Four Meals*. New York: Penguin Books.

Prétôt, L., Bshary, R., & Brosnan, S. F. (2016). Comparing species decisions in a dichotomous choice task: adjusting task parameters improves performance in monkeys. *Animal Cognition 19*(4), 819–834.

Raby, C. R., Alexis, D. M., Dickinson, A., & Clayton, N. S. (2007). Planning for the future by western scrub-jays. *Nature 445*(7130), 919–921.

Raby, C. R., & Clayton, N. S. (2009). Prospective cognition in animals. *Behavioural Processes 80*(3), 314–324.

Rachels, J. (1991). *Created From Animals*. Oxford University Press.

Raz, J. (1988). *The Morality of Freedom*. Oxford: Clarendon Press.

Redshaw, J. (2014). Does metarepresentation make human mental time travel unique? *Wiley Interdisciplinary Reviews: Cognitive Science 5*(5), 519–531.

Redshaw, J., & Bulley, A. (2018). Future Thinking in Animals: Capacities and Limits. In G. Oettingen, A. T. Sevincer, & P. M. Gollwitzer (Eds.), *The psychology of thinking about the future* (pp. 31–51). New York: Guilford. Authors' manuscript, final version.

Redshaw, J., & Suddendorf, T. (2013). Foresight beyond the very next event: four-year-olds can link past and deferred future episodes. *Frontiers in Psychology 4*, 404.

Redshaw, J., & Suddendorf, T. (2016). Children's and Apes' Preparatory Responses to Two Mutually Exclusive Possibilities. *Current Biology 26*(13), 1758–1762.

Redshaw, J., Taylor, A. H., & Suddendorf, T. (2017). Flexible Planning in Ravens? *Trends in Cognitive Sciences 21*(11), 821–822.

Regan, T. (1982). An Examination and Defense of One Argument Concerning Animal Rights. In *All That Dwell Therein: Animal Rights and Environmental Ethics* (pp. 113–147). Berkeley: University of California Press.

Regan, T. (2001). Animal Exploitation. In C. Cohen & T. Regan (Eds.): *The Animal Rights Debate* (pp. 135–149). Lanham: Rowman & Littlefield.

Regan, T. (2004). *The Case for Animal Rights*. Berkeley: University of California Press.

Regan, T. (2005). *Empty Cages: Facing the Challenge of Animal Rights*. Lanham: Rowman & Littlefield.

Regan, T. (2007). Die Tierrechtsdebatte. In Interdisziplinäre Arbeitsgruppe Tierethik (Ed.), *Tierrechte: Eine interdisziplinäre Herausforderung*. Erlangen: Harald Fischer Verlag.

Reinke, M. (2016). Mastschweine. In Albert Schweitzer Stiftung für unsere Mitwelt (Ed.). Retrieved from https://albert-schweitzer-stiftung.de/massentierhaltung/schweine/mastschweine (Nov 28, 2016).

Rinpoche, S. (2012). *The Tibetan Book of Living and Dying*. San Francisco.: HarperOne.

Roberts, W. A. (2012). Evidence for Future Cognition in Animals. *Learning and Motivation 43*(4), 169–180.

Roberts, W. A., & Feeney, M. C. (2009). The comparative study of mental time travel. *Trends in Cognitive Sciences 13*(6), 271–277.

Roberts, W. A., & Feeney, M. C. (2010). Temporal sequencing is essential to future planning: response to Osvath, Raby and Clayton. *Trends in Cognitive Sciences 14*(2), 52–53.

Rose, P. D. L. (2015). Warum Kinder keine Tierschlachtung sehen dürfen: Kindheits- und zivilisationstheoretische Anmerkungen. In G. Hirschfelder, A. Ploeger, J. Rückert-John, & G. Schönberger (Eds.), *Was der Mensch essen darf* (pp. 231–245). Wiesbaden: Springer Fachmedien.

Saidel, E. (2009). Attributing mental representations to animals. In R. W. Lurz (Ed.), *The Philosophy of Animal Minds* (pp. 35–51). Cambridge University Press.

Sauciuc, G.-A., Persson, T., Bååth, R., Bobrowicz, K., & Osvath, M. (2016). Affective forecasting in an orangutan: predicting the hedonic outcome of novel juice mixes. *Animal Cognition 19*(6), 1081–1092.

Scanlon, T. (1993). Value, Desire, and Quality of Life. In M. Nussbaum & A. Sen (Eds.), *The Quality of Life* (pp. 185–200). Oxford University Press.

Schacter, D. L., & Addis, D. R. (2007). The cognitive neuroscience of constructive memory: remembering the past and imagining the future. *Philosophical Transactions of the Royal Society B: Biological Sciences 362*(1481), 773–786.

Schacter, D. L., Benoit, R. G., & Szpunar, K. K. (2017). Episodic future thinking: mechanisms and functions. *Current Opinion in Behavioral Sciences 17*, 41–50.

Schechtman, M. (1994). The Same and the Same: Two Views of Psychological Continuity. *American Philosophical Quarterly 31*(3), 199–212.

Scholnick, E. K., & Friedman, S. L. (1987). The planning construct in the psychological literature. In S. L. Friedman, E. K. Scholnick, & R. R. Cocking (Eds.), *Blueprints for thinking. The role of planning in cognitive development* (pp. 33–38). Cambridge University Press.

Schroeder, T. (2009). Desire. In E. N. Zalta (Ed.), *Stanford Encyclopedia of Philosophy* (Summer 2015). Retrieved from http://plato.stanford.edu/archives/sum2015/entries/desire/.

Schweitzer, A. (2008). *Die Ehrfurcht vor dem Leben*. Munich: Beck.

Schweitzer, A. (2011). *Ehrfurcht vor den Tieren*. Munich: Beck.

Seed, A. M., & Dickerson, K. L. (2016). Future Thinking: Children But Not Apes Consider Multiple Possibilities. *Current Biology 26*(13), R525–R527.

Segner, H. (2012). *Fish: Nociception and pain. A biological perspective.* Bern: Bundesamt für Bauten und Logistik.

Shallice, T. (1982). Specific Impairments of Planning. *Philosophical Transactions of the Royal Society B: Biological Sciences 298*(1089), 199–209.

Shettleworth, S. J. (2010a). Clever animals and killjoy explanations in comparative psychology. *Trends in Cognitive Sciences 14*(11), 477–481.

Shettleworth, S. J. (2010b). Cognition, Evolution, and Behavior. Oxford University Press.

Simmons, A. (2006). *In Defense of an Animal's Right to Life*. Dissertation. Bowling Green State University.

Simmons, A. (2010). Do Animals Have an Interest in Continued Life? *Environmental Ethics 31*(4), 375–392.

Singer, P. (2000). Xenotransplantation and Speciesism. In H. Kuhse & P. Singer (Eds.): *Bioethics: An Anthology* (pp. 412–417). Oxford: Blackwell.

Singer, P. (2009). *Animal Liberation: The Definitive Classic of the Animal Movement.* New York: Harper Perennial Modern Classics.

Singer, P. (2011). *Practical Ethics*. Cambridge University Press.

Singer, P. (2015). Afterword. In In P. Singer, T. Višak & R. Garner (Eds.), *The Ethics of Killing Animals* (pp. 229–235). Oxford University Press.

Slors, M. (1998). Two Conceptions of Psychological Continuity. *Philosophical Explorations 1*(1), 61–80.

Sober, E. (2012). Anthropomorphism, Parsimony, and Common Ancestry. *Mind and Language 27*(3), 229–238.

Solomon, S., Greenberg, J., & Pyszczynski, T. (2015). *The Worm at the Core: On the Role of Death in Life*. London: Penguin.

Suddendorf, T. (2013a). Mental time travel: continuities and discontinuities. *Trends in Cognitive Sciences 17*(4), 151–152.

Suddendorf, T. (2013b). *The Gap: The Science of What Separates Us from Other Animals.* New York: Basic Books.

Suddendorf, T. (2017). The Emergence of Episodic Foresight and Its Consequences. *Child Development Perspectives 11*(3), 191–195.

Suddendorf, T., & Busby, J. (2005). Making decisions with the future in mind: Developmental and comparative identification of mental time travel. *Learning and Motivation 36*(2), 110–125.

Suddendorf, T., & Corballis, M. C. (1997). Mental time travel and the evolution of the human mind. *Genetic, Social, and General Psychology Monographs 123*(2), 133–167.

Suddendorf, T., & Corballis, M. C. (2007). The evolution of foresight: What is mental time travel, and is it unique to humans? *The Behavioral and Brain Sciences 30*(3), 299–313–351.

Suddendorf, T., & Corballis, M. C. (2008). Episodic memory and mental time travel. In E. Dere, A. Easton, L. Nadel & J. P. Huston (Eds.), *Handbook of Behavioral Neuroscience* (pp. 31–42). Amsterdam: Elsevier.

Suddendorf, T., & Corballis, M. C. (2010). Behavioural evidence for mental time travel in nonhuman animals. *Behavioural Brain Research 215*(2), 292–298.

Suddendorf, T., Corballis, M. C., & Collier-Baker, E. (2009). How great is great ape foresight? *Animal Cognition 12*(5), 751–754.

Suddendorf, T., Crimston, J., & Redshaw, J. (2017). Preparatory responses to socially determined, mutually exclusive possibilities in chimpanzees and children. *Biology Letters 13*(6).

Suddendorf, T., Watson, K., Bogaart, M., & Redshaw, J. (2019). Preparation for certain and uncertain future outcomes in young children and three species of monkey. *Developmental Psychobiology*. Early view online version before inclusion in an issue.

Suits, D. B. (2001). Why Death Is Not Bad for the One Who Died. *American Philosophical Quarterly 38*(1), 69–84.

Suits, D. B. (2012). Death and Other Nothings. *The Philosophical Forum 43*(2), 215–230.

Szpunar, K. K. (2010). Episodic Future Thought: An Emerging Concept. *Perspectives on Psychological Science 5*(2), 142–162.

Szpunar, K. K., Spreng, R. N., & Schacter, D. L. (2014). A taxonomy of prospection: Introducing an organizational framework for future-oriented cognition. *Proceedings of the National Academy of Sciences of the United States of America 111*(52), 18414–18421.

Szpunar, K. K., Spreng, R. N., & Schacter, D. L. (2016). Toward a Taxonomy of Future Thinking. In K. Michaelian, S. B. Klein, & K. K. Szpunar (Eds.), *Seeing the Future: Theoretical Perspectives on Future-Oriented Mental Time Travel* (pp. 21–36). Oxford University Press.

Tanner, J. (2009). The Epistemic Irresponsibility of the Subjects-of-a-Life Account. *Between the Species 13*(9).

Taylor, A. H., Hunt, G. R., Holzhaider, J. C., & Gray, R. D. (2007). Spontaneous metatool use by New Caledonian crows. *Current Biology 17*(17), 1504–1507.

Tecwyn, E. C., Thorpe, S. K. S., & Chappell, J. (2013). A novel test of planning ability: great apes can plan step-by-step but not in advance of action. *Behavioural Processes 100*, 174–184.

Tetzlaff, M., & Rey, G. (2009). Systematicity and intentional realism in honeybee navigation. In R. W. Lurz (Ed.), *The Philosophy of Animal Minds* (pp. 72–88). Cambridge University Press.

Tew, T. E., & Macdonald, D. W. (1993). The effects of harvest on arable wood mice (Apodemus sylvaticus). *Biological Conservation 65*(3), 279–283.

Thom, J. M., & Clayton, N. S. (2014). No evidence of temporal preferences in caching by Western scrub-jays (Aphelocoma californica). *Behavioural Processes 103*, 173–179.

Tomasello, M., & Call, J. (1997). *Primate Cognition*. Oxford University Press.

Tulving, E. (1972). Episodic and semantic memory. In E. Tulving & W. Donaldson (Eds.), *Organization of Memory* (pp. 839–847). New York: Academic Press.

Tulving, E. (1985). Memory and consciousness. *Canadian Psychology/Psychologie canadienne 26*(1), 1–12.

Tulving, E. (2005). Episodic Memory and Autonoesis: Uniquely Human? In H. S. Terrace & J. Metcalfe (Eds.), *The Missing Link in Cognition: Origins of Self-Reflective Consciousness* (pp. 3–56). Oxford University Press.

Višak, T. (2015). Do Utilitarians Need to Accept the Replaceability Argument? In P. Singer, T. Višak & R. Garner (Eds.), *The Ethics of Killing Animals* (pp. 117–135). Oxford University Press.

Višak, T. (2017). Cross-Species Comparisons of Welfare. In A. Woodhall & G. Garmendia de Trindade (Eds.), *Ethical and Political Approaches to Non-Human Animal Issues: Towards an Undivided Future*. New York: Palgrave MacMillan. Author manuscript.

Waal, F. B. M. de, & Ferrari, P. F. (2010). Towards a Bottom-Up Perspective on Animal and Human Cognition. *Trends in Cognitive Sciences 14*(5), 201–207.

Waal, F. B. M. de. (2016). *Are We Smart Enough to Know How Smart Animals Are?* New York: Norton

Wild, M. (2010). *Tierphilosophie zur Einführung*. Hamburg: Junius.

Wild, M. (2012). *Fische: Kognition, Bewusstsein und Schmerz. Eine philosophische Perspektive.* Bern: Bundesamt für Bauten und Logistik.

Williams, B. (1973). The Makropulos case: reflections on the tedium of immortality. In *Problems of the Self* (pp. 82–100). Cambridge University Press.

Wilson, T. D., & Gilbert, D. T. (2003). Affective Forecasting. *Advances in Experimental Social Psychology 35*, 345–411.

Wimpenny, J. H., Weir, A. A. S., Clayton, L., Rutz, C., & Kacelnik, A. (2009). Cognitive Processes Associated with Sequential Tool Use in New Caledonian Crows. *PLOS ONE* 4(8), e6471, 1–16.

Woodward, J. (2011). A Philosopher Looks at Tool Use and Causal Understanding. In T. McCormack, C. Hoerl, & S. Butterfill (Eds.), *Tool Use and Causal Cognition* (pp. 18–50). Oxford University Press. Author manuscript (pp. 1–31). Retrieved from http://philsci-archive.pitt.edu/8653/ (February 23, 2018).

Yalom, I. (2011). *Staring At The Sun: Being at peace with your own mortality. Overcoming the Terror of Death*. London: Piatkus.

Zentall, T. R., Singer, R. A., & Stagner, J. P. (2008). Episodic-like memory: Pigeons can report location pecked when unexpectedly asked. *Behavioural Processes* 79(2), 93–98.